New Water Regimes

Special Issue Editors

Jacque Emel
Alida Cantor

MDPI • Basel • Beijing • Wuhan • Barcelona • Belgrade

MDPI

Special Issue Editors
Jacque Emel
Clark University
USA

Alida Cantor
Portland State University
USA

Editorial Office
MDPI
St. Alban-Anlage 66
Basel, Switzerland

This edition is a reprint of the Special Issue published online in the open access journal *Resources* (ISSN 2079-9276) in 2018 (available at: http://www.mdpi.com/journal/resources/special_issues/ new_water_regimes).

For citation purposes, cite each article independently as indicated on the article page online and as indicated below:

Lastname, F.M.; Lastname, F.M. Article title. *Journal Name* **Year**, *Article number*, page range.

First Editon 2018

ISBN 978-3-03842-963-0 (Pbk)
ISBN 978-3-03842-964-7 (PDF)

Table of Contents

About the Special Issue Editors

Jacque Emel is a Professor of Geography at Clark University's Graduate School of Geography in Worcester, Massachusetts. Her research interests include political ecologies of industrial animal production, animal geographies, feminist intersectionality, water resources, and post-anthropocentric environmental justice.

Alida Cantor is an Assistant Professor of Geography at Portland State University. Her research focuses on water resources management and water law and policy, political ecology and political economy, hydropolitics, critical legal geography, and environmental justice. She also studies food systems and other issues of resource management and human-environment geography.

Preface to "New Water Regimes"

In a time of increasingly unpredictable water supply under climate change, water managers are struggling to handle the challenges of water supply, allocation, and water quality. Water management challenges are not only issues of hydrology; they also involve complex legal and political dimensions. This volume uses a legal geography perspective to examine the role of law and legal systems in water management. The articles in this collection provide critiques of existing water management practices, and also explore potential alternatives.

This volume was inspired by research and travels through the American Southwest during the 2012-2016 drought. The struggles to manage water resources during this drought highlighted not only the region's vulnerabilities to current and future climate change, but the shortcomings in the existing legal system's ability to handle contemporary challenges of water supply and water management more broadly. Water allocation in the Western United States is based largely on 'first come, first serve' principles of water rights developed over a hundred years ago. The legal framework for water management in the region was created based on a certain set of assumptions about water's value: namely, namely, that the 'highest and best use' of water was to be put to use by colonial settlers for economic gain.

However, in recent decades, other political theories such as eco-feminism, posthumanism, post-anthropocentricism, and decolonialism have produced different ideas about water valuation and management that conflict with Ameri-Euro imperialist goals and practices. Additionally, newer scientific knowledge, in particular about climate change, has shed doubt upon the ability of existing water rights systems to meet users' needs in an equitable way. In this volume, we explore questions of the limitations and opportunities of water law from a broad range of theoretical perspectives.

The volume asks a number of important questions, such as: How can we re-imagine entrenched water laws inaugurated in a period when capitalists and engineers considered rivers and aquifers nothing but plumbing opportunities to transfer huge volumes of water to agriculture and urban centers, especially in arid or semi-arid areas such as the western United States? What should alternative legal structures and principles look like, and what must they take into consideration? What constitutional reforms and political resistance movements are occurring in countries of the Global South? More broadly, how can we re-imagine civilizations that live with newer knowledges and more pluralistic understandings regarding humans in natures?

A variety of themes of water governance are addressed in this volume, including water allocation, groundwater management, collaborative governance, drought planning, and water quality. The papers describe and analyze water issues and new ideas in multiple countries, including Australia, Ecuador, New Zealand, India, and the United States.

The legal infrastructures that shape water management systems can be just as entrenched and difficult to change as the concrete infrastructures of dams and canals. However, as this volume illustrates, alternative modes of thinking are possible. Recognizing existing alternatives and proposing new alternative legal structures and principles that take into consideration Indigenous rights and more pluralistic understandings regarding humans in natures remains a difficult but important area of work. The papers in this special issue represent an important contribution towards this goal.

Jacque Emel and Alida Cantor
Special Issue Editors

resources

MDPI

Editorial

New Water Regimes: An Editorial

Alida Cantor [1,*] and Jacque Emel [2]

[1] Department of Geography, Portland State University, 1721 SW Broadway, Portland, OR 97201, USA
[2] Graduate School of Geography, Clark University, 950 Main Street, Worcester, MA 01610, USA;
     jemel@clarku.edu
* Correspondence: acantor@pdx.edu; Tel.: +1-503-725-3165

Received: 22 March 2018; Accepted: 2 April 2018; Published: 4 April 2018

Abstract: This editorial is an introduction to the special issue of *Resources* on New Water Regimes. The special issue explores legal geographies of water resource management with the dual goals of providing critiques of existing water management practices as well as exploring potential alternatives. The papers in the special issue draw from numerous theoretical perspectives, including decolonial and post-anthropocentric approaches to water governance; social and environmental justice in water management; and understanding legal ecologies. A variety of themes of water governance are addressed, including water allocation, groundwater management, collaborative governance, drought planning, and water quality. The papers describe and analyze water issues and new ideas in multiple countries, including Australia, Ecuador, New Zealand, India, and the United States.

Keywords: water resource management; water rights; water quality; legal geography; political ecology

1. Introduction

In this special issue, we critically examine the legal and administrative structures of water control, with the goal of imagining alternatives to entrenched systems of capitalist and anthropocentric water governance. While concrete water infrastructure projects such as dams are readily recognized as difficult to change, the laws and legal principles that shape water management can represent equally deep-rooted structures that are just as hard to shift. However, as successful dam removal projects illustrate, even concrete is not immutable; similarly, legal structures can also change and evolve. This issue focuses on developing a better understanding of how existing laws and administrative structures shape water management, often in ways that preclude more sustainable and equitable practices, while also considering alternatives informed by perspectives such as post-anthropocentrism, decolonialism, and social and environmental justice.

The quantity and quality of water resources throughout the world are increasingly stressed by many intersecting factors, including climate change, urban and population growth, and economic and industrial development. The particular dimensions of the many struggles to manage water sustainably and equitably differ widely from region to region. However, a common thread linking water resource management challenges around the world is that these challenges are not just hydrological ones—they are socio-legal in nature as well.

2. Critical Legal Geographies of Water Resource Management

The socio-legal dimensions of water management challenges are readily evident; solutions are less forthcoming. For example, the struggles to manage the severe drought facing the Western United States from 2012 to 2016 sharply highlighted not only the region's socio-ecological vulnerability to future climate change, but also shortcomings in the abilities of the existing legal system to handle contemporary challenges of water supply and allocation. Water allocation practices in this region, namely the prior appropriation system, are based largely on a century-old set of assumptions about water valuation and management: that the 'highest and best use' of water was to be put to use by colonial settlers for economic gain, and that water users must continually put their water to 'productive' use or risk losing their right. Today, this legal water rights system arguably hampers the ability to manage water in congruence with contemporary understandings and realities of climate change in order to meet the diverse needs of human and non-human water users. Yet, changing the system is by all measures nearly unimaginable; proposals for how to adapt to climate change and contemporary water needs typically assume the immutability of this system and are limited to imagining strategies that work within the existing water rights system.

Outdated water rights systems that limit adaptive capacity in the face of socio-ecological change are not the only problems facing water governance today. Unequal access to adequate water supplies due to degraded quality, limited quantity, and/or high cost remains a problem for many water users. Legal systems that have evolved to separately manage water quality and quantity, as well as groundwater and surface water, also present barriers to integrated and sustainable water management practices. The rights of Indigenous communities, as well as the rights of nonhuman water users, have received woefully inadequate consideration in many systems of water governance as currently practiced. Water management remains deeply rooted in capitalist, colonialist, and imperialist structures.

Several areas of scholarship contribute to a deeper understanding of these critiques. In particular, scholarship on critical legal geographies has sought to develop a better understanding of the interactions between law and place/space [1,2]. Legal geography, which starts from the premise that law, society, and space are co-constituted, has expanded significantly in recent years [2,3]. Of particular interest to this special issue, recent scholarship within legal geography has begun to examine the co-production of law and more-than-human environments [3]. Meanwhile, an often-separate literature on the "hydrosocial cycle" has examined political economies and ecologies of water resource governance [4,5]. Within this literature, scholars have developed critiques of modernist water resource management, which privileges scientific hydrologic expertise and concentrates control and management of water in agencies of the state [4–8]. Recently, scholars have taken up the challenge of merging the insights of legal geography with those of hydrosocial perspectives, examining water law(s) from a critical geographic perspective to better understand the socio-ecologies and socio-environmental injustices produced through interactions between multiple legal and hydro-social systems [9–12]. This literature emphasizes that water law is not a monolith: often, multiple overlapping socio-legal orders may exist simultaneously [12–14].

In this issue, we build upon this work, which aims to critically understand practices in modern water management, focusing on the intersections between water governance and law. We also argue that not only is it possible, it is necessary to imagine different futures for water governance. In recent decades, newer political theories such as eco-feminism, posthumanism, post-anthropocentricism, and decolonialism have produced different ideas about water valuation and management that challenge capitalist and imperialist goals and practices. This issue engages a variety of theoretical perspectives in critically examining water governance.

3. Contributions to the Special Issue

The papers in this special issue build upon this scholarship by providing a theoretical and empirical basis for change toward a more sustainable, less anthropocentric form of water governance, while also outlining the many barriers and constraints that stand in the way of actually attaining this kind of water governance. The papers describe and analyze water issues and new ideas in several countries, including Australia, Ecuador, New Zealand, India, and the United States. Several of the papers examine water issues in California, the biggest user of water in the United States and a place with continuing supply and quality problems. By utilizing detailed empirical case studies, the papers in this special issue examine specific dimensions of opportunities and constraints for new water regimes rooted in particular places. In this introduction, we discuss the contributions of the articles in this special issue in light of several theoretical threads, including non- or post-anthropocentrism, post- or de-colonialism, social and environmental justice, and legal ecologies. In addition to describing the many challenges and barriers that are discussed in the articles, we attempt to highlight opportunities and points of hope within each piece.

3.1. Post-Anthropocentric Approaches to Water Governance

Post-anthropocentrism, a philosophical stance toward the Earth based on ecofeminism, animal philosophy, deep ecology, Indigenous thought, and certain strands of poststructuralism, envisages a world or worlds that are not human-centered but multi-centered. In these worlds or ontologies, beings other than humans may have legal standing [15]. Several of the pieces in this issue, including those by Jamie McEvoy et al. and Lidia Cano Pecharroman, specifically consider non-anthropocentric perspectives on water laws and management practices. As the authors of both papers claim, such recognition is slow to arrive and complex in administration. Yet both provide some optimism that a more-than-human approach to water governance is indeed possible.

In their article on "Ecological Drought: Accounting for the Non-Human Impacts of Water Shortage in the Upper Missouri Headwaters Basin, Montana, USA", McEvoy et al. write of the failure of U.S. water law to recognize water requirements for non-humans (both animals and plants), explaining that most drought planning efforts focus primarily on human water users. Where nonhuman water users are considered, they are given consideration in certain anthropocentric ways; the authors conclude that most practices in the watersheds of the Upper Missouri in Montana focus on instream flows and temperatures for certain species of sport fish. This paper illustrates the strong influence of commodity orientations toward water resources, even though instream flow requirements, which have been in place in Montana for over four decades, purportedly recognize more-than-human water needs. The authors suggest that despite these criticisms, drought plans do provide a potential starting point for recognizing the ecological impacts of drought, and that with a deeper consideration of ecological processes, a more-than-human orientation toward drought is indeed possible.

In "Rights of Nature: Rivers that Can Stand in Court", Pecharroman provides a hopeful perspective, discussing recent legal cases from multiple countries that give legal rights to rivers. In the article, she discusses numerous legal and philosophical groundings behind granting legal rights to nature, including several examples of Indigenous perspectives. Pecharroman then describes case studies that have emerged from New Zealand, Ecuador, India, and Colombia within the past five years, where mostly tribal or Indigenous groups have fought successfully in court for the recognition of rivers to possess legal rights. In the paper Pecharroman suggests that while legal recognition of the rights of nature is still in very early stages, and many uncertainties still exist, we are arguably witnessing an important process of legal confirmation of ecological values, as non-human subjects are granted legal rights.

3.2. Decolonial Water Governance and Legal Pluralism

What is sometimes termed decolonial water governance is actually non-European governance systems that are being asserted in the global south and in settler colonies. These so-called "worlding" systems of governance do not follow the European legal and infrastructure legacies imprinted across the imperialist track [16]. Negotiations between settler-colonial and Indigenous groups frequently reveal problems of "translation" or incommensurability as different ontological perspectives on water prove to be incompatible [17,18]. In addition, water laws are frequently layered strata [19], with pre-European systems and settler-colonial laws coexisting sometimes uneasily on top of one another. An approach of "legal pluralisms" [12,20] acknowledges the existence of these multiple-layered legal systems, even though the negotiations can, in practice, be difficult.

Lana Hartwig, Sue Jackson, and Natalie Osborne's article describes the challenges of working towards decolonial water governance in a settler-colonial state. The article "Recognition of Barkandji

Water Rights in Australian Settler-Colonial Water Regimes" describes the attempts of Indigenous peoples in what is now called Australia to have their water rights recognized by Australian governments. As the paper describes, the river in question is central to the existence of the Barkandji People. However, recognition has been a long and complicated process given the extent of historical injustices, not only because of the explicit "loss" perceived by settler-colonists in "giving up" what they've actually stolen, but because ontologies collide in the existing legal system which identifies "water" as a particular economic artefact as opposed to the multidimensional and subject-centered conceptualization of water held by the Barkandji. Hartwig et al. describe the ways in which the misrecognition (i.e., oversimplified, restricted, overlooked, and stereotyped) or even altogether non-recognition of Aboriginal water rights perpetuates the status quo of colonial power. However, at the same time, they argue that in so doing, the legitimacy of state water regimes is actually undermined, as the state fails to generate genuine respect. Thus, there potentially exists a genuine motivation for settler-colonial governments to genuinely recognize Indigenous rights (and to consult Indigenous people meaningfully in the process).

3.3. Justice and Equity

While much scholarship and activism in terms of environmental justice and water focuses primarily on water quality concerns [21], the issue of equity in water allocation is also a long-standing concern of many water scholars [19,22]. Several papers in this special issue, including those by Ann Drevno, by Zachary Sugg, and by Andrés Martínez Moscoso, Víctor Gerardo Aguilar Feijó and Teodoro Verdugo Silva, focus specifically on justice and equity concerns related to water supply allocation as well as water quality. Sugg and Moscoso, Feijó and Silva focus on equitable water allocation, while Drevno's article focuses on water quality. All three of these papers speak to the existence of gaps between existing laws and equitable on-the-ground outcomes; they also point to the need for more nuanced and locally-sensitive systems in place of overly prescriptive approaches.

Sugg, in "An Equity Autopsy: Exploring the Role of Water Rights in Water Allocations and Impacts for the Central Valley Project during the 2012–2016 California Drought", argues that strict adherence to a priority system for water allocations produces inequitable socio-ecological outcomes during drought. In California's recent drought, thousands of water users had their water rights curtailed, with highly uneven socioeconomic and ecological impacts. Sugg points out that the current water governance system is too laden with conflict to be considered effective or well-adapted. The concept of equitable apportionment, which involves allocating water according to multiple relevant criteria beyond simply a priority system, is recommended as a more just way of allocating water in drought.

In their paper on "The Vital Minimum Amount of Drinking Water Required in Ecuador", Moscoso, Feijó and Silva illustrate the difficulties of implementing a new approach to water allocation in Ecuador. Ecuador recently recognized water as a fundamental human right, and in doing so, established a minimum quantity of water that is guaranteed as a constitutional right. However, this raises questions of how much water is guaranteed, and who will be responsible for paying for the water. The authors find that the effort to ensure water for all can have counterintuitive results, and may even widen gaps in equity, as inefficient water providers transfer the costs of providing services to those who are already economically vulnerable. They note that the water provisioning system must take better account of regional variations in order to avoid this pitfall.

Drevno, focusing on water quality, also speaks to the need to take into account local specificities in water management practices. Drevno's article, "From Fragmented to Joint Responsibilities: Barriers and Opportunities for Adaptive Water Quality Governance in California's Urban-Agricultural Interface", discusses the consequences of managing urban and agricultural water quality separately in a location where urban development and agricultural lands are in close proximity to one another. By examining a case study at the urban-agricultural interface, she illustrates the problems with California's current system addressing urban water quality and agricultural water quality under different systems. Drevno calls for agricultural water quality, in particular, to be addressed more rigorously, and for the two systems to be aligned more closely in order to alleviate environmental justice concerns of poor water quality.

3.4. Legal Ecologies

A final theme of this special issue is that of legal ecologies, which we define as the specific places and more-than-human ecosystems that are co-produced along with laws, contracts, and their implementation. The ecologies of water law include the ecosystems and species impacted by the transfer of water from one place to another, along with the new ecologies produced in urbanized and irrigated areas.

Julia Sizek's communication on "California's Groundwater Regime: The Cadiz Case" discusses a proposed groundwater extraction project that would transfer water from the desert of Southeastern California to urban Southern California, with potentially dramatic impacts on fragile desert ecosystems. The Cadiz project is a notorious one, an idea that has long been considered a failure, only to be resurrected under the Trump administration. Sizek describes the underlying politics behind the project, including the role of hotly contested science in justifying/challenging the project. Sizek demonstrates that in California, groundwater law is fundamentally linked to private property ownership of land, meaning that legal ecologies are, in this case, not only shaped by water laws but by landed property regimes—themselves an artefact of settler-colonial states—as well.

4. Concluding Remarks

In conclusion, nearly all of the papers in this special issue illustrate the difficulties of actually implementing new visions of water governance. They demonstrate the powerful continuity and hegemony of certain legal systems, particularly in the settler-colonial legal systems rooted in European traditions. Yet they also provide examples of how water laws are changing—or could change—and are being pushed in new directions that recognize the rights of Indigenous people, the rights of nature, and the values of more just, less anthropocentric, and more integrated systems for water management.

Looking forward, we encourage critical scholars to continue to consider legal and administrative systems of water governance from a broad range of theoretical perspectives. Seeking to transform the entrenched water laws that were inaugurated in a period when capitalists and engineers considered rivers and aquifers nothing but plumbing opportunities to transfer huge volumes of water to agriculture and urban centers is not a light task. Recognizing existing alternatives and proposing new alternative legal structures and principles that take into consideration Indigenous rights and more pluralistic understandings regarding humans in natures remains a difficult but important area of work. Critical scholars can propose legal reforms that go beyond a contemporary fixation with market-based imperatives as the main solution for prior appropriation water allocation systems. They can also examine the intersection of water law and governance with other related important issues, such as climate change, food and water security, and food and water sovereignty. Comparative studies can contribute to constitutional reform and political resistance around the globe. The papers in this special issue represent an important contribution, yet there remains much to be done, theoretically, empirically, and practically.

Acknowledgments: Thanks to all of the contributors to the special issue, as well as to the anonymous reviewers who have contributed to the development of the articles. No particular funding source was associated with this work.

Author Contributions: Alida Cantor and Jacque Emel both contributed in equal measure to recruiting and reviewing papers for this special issue and to writing this editorial.

Conflicts of Interest: The authors declare no conflict of interest.

References

1. Braverman, I.; Blomley, N.K.; Delaney, D.; Kedar, A. *The Expanding Spaces of Law: A Timely Legal Geography*; Stanford Law Books: Stanford, CA, USA, 2014; ISBN 9780804791878.
2. Delaney, D. Legal geography I: Constitutivities, complexities, and contingencies. *Prog. Hum. Geogr.* **2015**, *39*, 96–102. [CrossRef]
3. Delaney, D. Legal geography III. *Prog. Hum. Geogr.* **2017**, *41*, 667–675. [CrossRef]
4. Linton, J.; Budds, J. The hydrosocial cycle: Defining and mobilizing a relational-dialectical approach to water. *Geoforum* **2014**, *57*, 170–180. [CrossRef]
5. Swyngedouw, E. The Political Economy and Political Ecology of the Hydro-Social Cycle. *J. Contemp. Water Res. Educ.* **2009**, *142*, 56–60. [CrossRef]
6. Linton, J. *What Is Water? The History of a Modern Abstraction*; UBC Press: Vancouver, BC, Canada, 2010; ISBN 0774817038.
7. Linton, J. Modern water and its discontents: A history of hydrosocial renewal. *Wiley Interdiscip. Rev. Water* **2014**, *1*, 111–120. [CrossRef]
8. Swyngedouw, E. Modernity and hybridity: Nature, regeneracionismo, and the production of the Spanish waterscape, 1890–1930. *Ann. Assoc. Am. Geogr.* **1999**, *89*, 443–465. [CrossRef]
9. Cantor, A. The public trust doctrine and critical legal geographies of water in California. *Geoforum* **2016**, *72*, 49–57. [CrossRef]
10. Perramond, E.P. Water governance in New Mexico: Adjudication, law, and geography. *Geoforum* **2013**, *45*, 83–93. [CrossRef]
11. Jepson, W. Claiming Space, Claiming Water: Contested Legal Geographies of Water in South Texas. *Ann. Assoc. Am. Geogr.* **2012**, *102*, 614–631. [CrossRef]
12. Boelens, R.; Vos, J. Legal pluralism, hydraulic property creation and sustainability: The materialized nature of water rights in user-managed systems. *Curr. Opin. Environ. Sustain.* **2014**, *11*, 55–62. [CrossRef]
13. Roth, D.; Boelens, R.; Zwarteveen, M. *Liquid Relations: Contested Water Rights and Legal Complexity*; Rutgers University Press: New Brunswick, NJ, USA, 2005; ISBN 0813536758.
14. De vos, H.; Boelens, R.; Bustamante, R. Formal law and local water control in the Andean region: A fiercely contested field. *Water Resour. Dev.* **2006**, *22*, 37–48. [CrossRef]
15. Stone, C.D. Should Trees Have Standing—Toward Legal Rights for Natural Objects. *S. Calif. Law Rev.* **1972**, *45*, 450.
16. Furlong, K.; Kooy, M. Worlding Water Supply: Thinking Beyond the Network in Jakarta. *Int. J. Urban Reg. Res.* **2017**, *41*, 888–903. [CrossRef]
17. Nirmal, P. *Disembodiment and Deworlding: Taking Decolonial Feminist Political Ecology to Ground in Attappady, Kerala*; Clark University: Worcester, MA, USA, 2017.
18. Povinelli, E.A. *The Cunning of Recognition: Indigenous Alterities and the Making of Australian Multiculturalism*; Duke University Press: Durham, NC, USA, 2002; ISBN 0822328682.
19. Brown, F.L.; Ingram, H.M. *Water and Poverty in the Southwest*; University of Arizona Press: Tucson, AZ, USA, 1987.
20. Robinson, D.F.; Graham, N. Legal pluralisms, justice and spatial conflicts: New directions in legal geography. *Geogr. J.* **2018**, *184*, 3–7. [CrossRef]
21. Vanderwarker, A. Water and environmental justice. In *A Twenty-First Century US Water Policy*; Christian-Smith, J., Gleick, P.H., Cooley, H., Allen, L., Vanderwarker, A., Berry, K., Eds.; Oxford University Press: New York, NY, USA, 2012; p. 52.
22. Emel, J. Resource instrumentalism, privatization, and commodification. *Urban Geogr.* **1990**, *11*, 527–547. [CrossRef]

resources

MDPI

Article

Ecological Drought: Accounting for the Non-Human Impacts of Water Shortage in the Upper Missouri Headwaters Basin, Montana, USA

Jamie McEvoy [1,*], Deborah J. Bathke [2], Nina Burkardt [3], Amanda E. Cravens [3], Tonya Haigh [2], Kimberly R. Hall [4], Michael J. Hayes [5], Theresa Jedd [2], Markéta Poděbradská [2] and Elliot Wickham [2]

[1] Department of Earth Sciences, Montana State University, Bozeman, MT 59715, USA
[2] National Drought Mitigation Center and School of Natural Resources, University of Nebraska, Lincoln, NE 68583, USA; dbathke2@unl.edu (D.J.B.); thaigh2@unl.edu (T.H.); tjedd2@unl.edu (T.J.); podebradska.marketa@huskers.unl.edu (M.P.); Elliot.wickham@huskers.unl.edu (E.W.)
[3] Fort Collins Science Center, U.S. Geological Survey, Fort Collins, CO 80526, USA; burkardtn@usgs.gov (N.B.); aecravens@usgs.gov (A.E.C.)
[4] North America Region, The Nature Conservancy, Haslett, MI 48840, USA; kimberly.hall@tnc.org
[5] School of Natural Resources, University of Nebraska, Lincoln, NE 68583, USA; mhayes2@unl.edu
* Correspondence: jamie.mcevoy@montana.edu; Tel.: +1-406-994-4069

Received: 18 December 2017; Accepted: 14 February 2018; Published: 20 February 2018

Abstract: Water laws and drought plans are used to prioritize and allocate scarce water resources. Both have historically been human-centric, failing to account for non-human water needs. In this paper, we examine the development of instream flow legislation and the evolution of drought planning to highlight the growing concern for the non-human impacts of water scarcity. Utilizing a new framework for ecological drought, we analyzed five watershed-scale drought plans in southwestern Montana, USA to understand if, and how, the ecological impacts of drought are currently being assessed. We found that while these plans do account for some ecological impacts, it is primarily through the narrow lens of impacts to fish as measured by water temperature and streamflow. The latter is typically based on the same ecological principles used to determine instream flow requirements. We also found that other resource plans in the same watersheds (e.g., Watershed Restoration Plans, Bureau of Land Management (BLM) Watershed Assessments or United States Forest Service (USFS) Forest Plans) identify a broader range of ecological drought risks. Given limited resources and the potential for mutual benefits and synergies, we suggest greater integration between various planning processes could result in a more holistic consideration of water needs and uses across the landscape.

Keywords: ecological drought; drought planning; prior appropriation; instream flows; Upper Missouri Headwaters Basin; Montana

1. Introduction

1.1. Overview

Water laws and drought plans are used to prioritize and allocate scarce water resources. Both have historically been human-centric, failing to account for non-human water needs. We begin this paper with an overview of water laws that govern water allocation in the western U.S. and the development of instream flow legislation to redress some of the ecological impacts caused by early water development. We then provide a brief history of drought planning, which has primarily focused on the agricultural and socio-economic impacts of drought. Utilizing a new framework for ecological drought, we introduce the case study and methods used to understand if, and how, the ecological impacts of drought are currently being assessed in seven watersheds in southwestern Montana (MT). After presenting the results of our analysis, we discuss the limitations of legislation and drought plans focused on minimum streamflows and suggest opportunities for more comprehensive approaches to watershed drought planning.

1.2. Water Rights in the Western U.S.

Often described as "first in time, first in right," the legal doctrine of prior appropriation that governs water allocation throughout most of the western U.S. is a system that gives those who first made use of water ("first in time") higher priority rights to the water ("first in right"). Unlike the common law system used in the eastern U.S. that tied rights to water to land ownership of land bordering a river, the prior appropriation system allowed users to divert water to properties away from the river, a key development in an arid environment. Prior appropriation assumes that someone is diverting water away from a stream to put it to a "beneficial use." The original 19th century notion was that beneficial uses included agriculture, mining, domestic use, and other industry; water left in the river was presumed to be available for someone else to divert and claim [1,2].

The prior appropriation doctrine has three important implications when water becomes scarce. First, in times of water shortage, the user with the most senior right to divert water ("water rights") receives their full allotment before the next user receives any water. Second, prior appropriation contains a provision (originally designed to prevent speculation and hording of water) that if a water right is not used regularly, it is considered abandoned; fear of this "use it or lose it" rule provides a disincentive for keeping water in the stream. Finally, if the number of water rights on a given section of river becomes great enough, the human needs for which water is being diverted (e.g., agriculture, mining, urban development) and the needs of species and ecosystems that depend on water being left in the river can come into conflict. Even in a "normal" year, there can be insufficient water left in the river for fish and other ecosystem needs during later summer months, especially in places where stream flows are fed by snowmelt. The scarcity of water during times of meteorological drought (i.e., dry weather patterns and below average precipitation) amplifies these tensions. In southwestern Montana, water shortages have led to the adoption of innovative new mechanisms aimed at keeping more water instream.

1.3. Mechanisms for Protecting Instream Environmental Flows

The early "beneficial uses" of water in the western U.S. were consumptive uses that diverted water away from a stream (e.g., mining, irrigation, municipal water use). With increasing recognition of the benefits of adequate water quantity and quality to ecological and human systems, state and federal legislation has more recently focused on protecting instream flow values. Federal laws such as the Wild and Scenic Rivers Act of 1968 [3] and section 404 of the Clean Water Act [4] provided instream flow protection for defined purposes and specific reaches of streams or rivers. The first state to give legal recognition to instream uses of water was Oregon, in 1955. Montana took steps toward instream flow protection in 1969 by establishing "Murphy Rights" for twelve Montana streams. The Montana Fish and Game Commission can file to allocate unappropriated water in the amount necessary to protect fish and wildlife habitat on these blue-ribbon streams [5].

The Montana Water Use Act of 1973 [6] established a mechanism for state and federal entities to seek limited instream flow protection; these rights are held by Montana Fish, Wildlife, and Parks (MFWP). The allowable purposes of an instream flow are broad in Montana and include protection of existing or future beneficial uses, or maintenance of minimum flow, water quality or water level. When an existing consumptive water right is changed to an instream flow use, the original date of the water right applies. For example, in the Blackfoot drainage public recreation water rights exist with dates as far back as 1928, while public recreation claims in the Bitterroot basin are quite junior, with priority dates in the 1970s.

Minimum flow is usually a single threshold beyond which water cannot be withdrawn for consumptive use, and frequently provides sub-optimal habitat conditions for aquatic species [7]. The "wetted-perimeter" method, which calculates the amount of water needed to sustain critical habitat for fish migration, spawning and food production at the shallowest part a river (usually a riffle) is one of the more common ways of quantifying the instream flow needs of fish [8]. As scientific understanding of the dynamic nature of river ecosystems has increased, some laws and regulations have shifted to support variable flow regimes, rather than minimum flows. Processes for establishing instream flow recommendations, especially more comprehensive methods, can be costly, time-intensive, and require specialized knowledge [9].

Limitations of instream flow policies in Montana include: (1) Instream flows are bounded by original 'dates of use,' making it difficult to transfer to use for a specific time of year; (2) even the oldest instream flow rights are quite junior and are subject to the "first in time is first in right" rule. As an example, Murphy rights are junior in priority and can be revoked if a District Court determines the water should be put to use in a way that is more beneficial to the public; (3) instream flow rights are subject to reviews and can be modified or revoked if the objectives of the reservation are not being met [10], if another applicant is deemed to be a "qualified reservant" [11], or if the reservation may harm future development [12]; (4) rights cannot exceed 50% of the annual flow of the stream, which may not be adequate to protect some values [12].

Another set of mechanisms for maintaining instream flows include cooperative agreements and drought planning. In some cases, the threat of an endangered species listing provides an incentive for water users to proactively develop a plan to mitigate the impacts of water shortage through a Candidate Conservation Agreement with Assurances (CCAA). These agreements between the U.S. Fish and Wildlife Service (USFWS) and any non-Federal entity allow the latter to voluntarily implement conservation activities that remove threats to endangered or threatened species in exchange for assurance that, if the species were to become listed under the Endangered Species Act (ESA), the landowner or water user would not be subject to additional restrictions [13]. In the Upper Missouri Headwaters (UMH), the fluvial artic grayling (*Thymallus arcticus*) is the target species for CCAAs which aim to:

1. Improve streamflows;
2. Improve and protect the function of riparian habitats;
3. Identify and reduce or eliminate entrainment threats for grayling;
4. Remove barriers to grayling migration [14].

Beyond these contractual agreements, there are also examples of extralegal "shared sacrifice" agreements [15,16] whereby neighbors agree to reduce irrigation withdrawals in dry years, typically based on a predetermined streamflow level and/or water temperature that triggers a reduction in irrigation withdrawals. In some cases, these "shared-sacrifice" agreements are very informal—based only a handshake. In other cases, these agreements are made legible through a written and publically-available document. In the UMH basin, these documents range from a 1988 *Plan to Avoid Dewatering of the Ruby River Project* [17] that is primarily focused on ensuring that senior water rights holders receive their full allotments in dry years to a recent 2016 *Beaverhead Watershed Drought Resiliency Plan* [18] that provides a detailed assessment of the regional geography and climate, a relatively broad list of drought impacts, and an overview of current and potential drought indicators. As discussed below, the development of watershed drought plans is a priority for the state of Montana.

Other legal mechanisms for protecting instream flows in Montana include leasing or conversions, authorized by the Water Leasing Bill (HB 707; 1989); purchase or service contracts for water delivery from water storage projects, and designation of closed basin watersheds. However, it is beyond the scope of this paper to discuss these mechanisms.

1.4. History of Drought Planning

Droughts in the United States have a long legacy of devastating agricultural impacts that have affected crop production and disrupted settlement patterns and livelihoods [19]. The quantification of agricultural impacts is still the main source of additions to the list of 25 "Billion Dollar" drought disasters across the U.S. since 1980 [20]. Recently, however, there is a growing awareness of the complex range of drought impacts that affect a broad spectrum of sectors including natural resources, energy, recreation and tourism, and public health [21,22]. Although less apt to be quantified [23], the impacts of droughts on non-agricultural livelihoods and ecosystems are considerable. Past droughts impacts, and the projections that future droughts may become more frequent and severe in some parts of the country [24,25], illustrate that planning ahead for drought events at a variety of scales and sectors is essential.

Drought planning in the U.S. is a relatively recent phenomenon that has lagged behind the planning for other natural hazards. It first appeared as a concept when Don Wilhite, at the University of Nebraska, analyzed the federal response to the severe 1976–77 drought event and began advocating for proactive drought risk management at the state and local levels [26]. Three key elements of drought risk management include: (1) early warnings of drought risk, based on the continuous assessment of appropriate indicators, (2) impact and vulnerability assessment, and (3) response and mitigation strategies [26,27]. Accurate early warning information is essential for drought risk management because decision makers require this information to implement effective drought response activities, recovery programs, and proactive drought policies [28]. Impact and vulnerability assessments identify the principal activities, groups, or regions most at risk and why. The response and mitigation component should identify both actions that minimize impacts during a drought event (response strategies) and actions that reduce risk in advance of drought (mitigation strategies). While drought impacts are increasingly recognized across multiple sectors [21], incorporating the full suite of sectors into drought planning is a challenge. In particular, the ecological impacts of drought, and the connections between natural systems, ecosystem services, and societal values (e.g., non-agricultural livelihoods, public health, recreational opportunities) have not been accounted for in most drought plans. Case studies from across the globe, including California [29,30], Australia [31,32], Kenya [33], and Bangladesh [34] have brought attention to the need for assessing and understanding the ecological impacts of drought.

Following a recent publication by Crausbay and Ramirez et al., highlighting the importance of considering risks to people and nature from ecological drought impacts [35], we investigated the extent to which ecological risks are already being incorporated into drought planning in the Upper Missouri Headwaters (UMH) region in southwestern Montana. Ecological drought is defined as "an episodic deficit in water availability that drives ecosystems beyond thresholds of vulnerability, impacts ecosystem services, and triggers feedbacks in natural and/or human systems" [35] (p. 24). Crausbay and Ramirez et al. [35] pair this definition with a framework (Figure 1) that promotes a holistic consideration of drought in which natural processes (e.g., precipitation, transpiration by plants) and human actions (e.g., water use, land management that interact with hydrologic processes) interact to influence drought impacts [21,36]. The premise behind this definition and framework is that a more complete accounting of what influences water availability and drought risks to both ecosystems and people will lead to more effective plans and more sustainable communities (see also [37,38]).

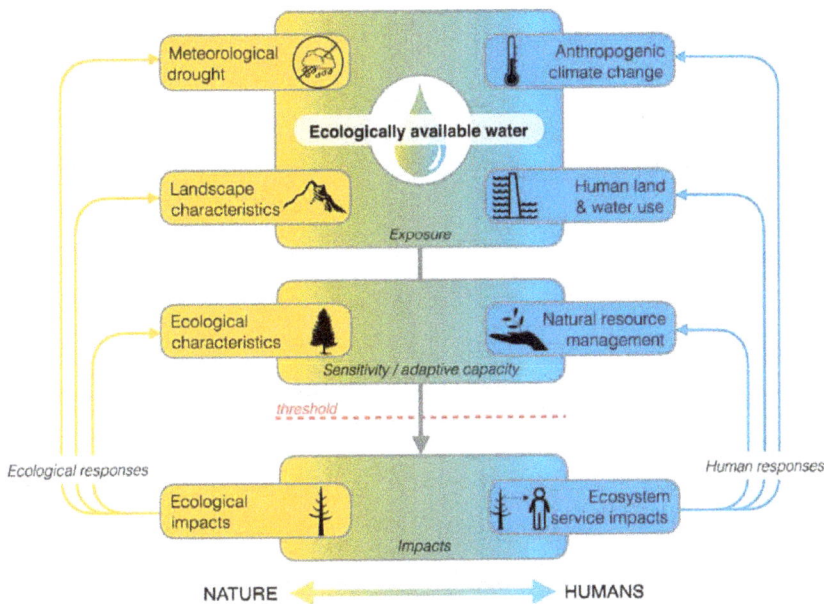

Figure 1. Ecological drought framework from Crausbay and Ramirez et al. [35].

2. Case Study Description

Drought preparedness and planning are named as priorities in Montana's 2015 State Water Plan, which emphasizes that:

> Drought preparedness requires a collaborative approach within small- to medium- sized watersheds. Working together, water users and water management agencies can develop adaptive management strategies that can yield benefits to water supply, fisheries, and water quality. Adaptive management also requires effective coordination between state and federal agencies responsible for managing water supply, water quality, fisheries, and drought and water supply forecasting. [39] (p. 69)

One region within Montana that has been the particular focus of such an approach is the Upper Missouri Headwaters (UMH) in southwestern Montana (i.e., Missouri Headwaters HUC 6 Watershed; Figure 2). The UMH was selected as one of two demonstration projects by the Obama Administration's National Drought Resilience Partnership (NDRP). The State of Montana and the NDRP—a "collaborative of federal and state agencies, non-governmental organizations (NGOs), and watershed stakeholders–are working together to leverage and deliver technical, human and financial resources to help address drought in the arid West" and particularly in the UMH [40] (p. 2). The UMH region has experienced frequent droughts, changing land use practices, and population growth. The landscape holds a variety of ecological values, including connectivity within the northern Rockies ecoregion, that are important to a variety of species and human stakeholders. The NDRP project focuses on providing state and federal resources to meet the goal of "empowering local communities to prepare for and mitigate the impacts of drought on livelihoods and the economy" [40] (p. 2). It thus builds on Montana's strong legacy of local watershed groups working to restore and protect watersheds [41] both within the UMH and farther afield; these groups have pioneered local-scale solutions to drought preparedness, including "shared sacrifice" agreements between farmers and anglers [40,41].

Study Area for Local Level (HUC 8 Watershed) Drought Planning Efforts

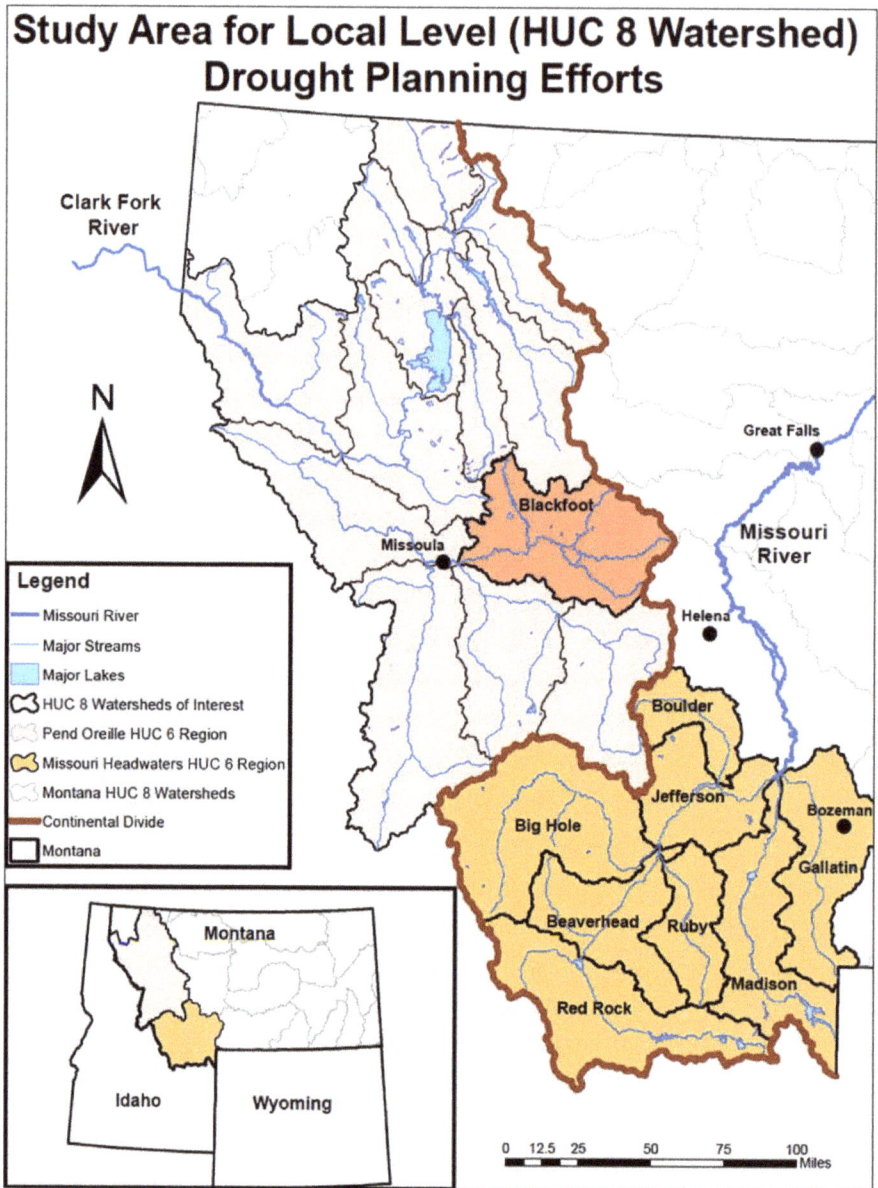

Figure 2. Map highlighting watersheds in the Upper Missouri Basin, and the Blackfoot watershed.

This study focuses on the six watersheds of the UMH basin (Boulder-Jefferson, Big Hole, Gallatin, Madison, Ruby, and Beaverhead-Red Rock), as well as the Blackfoot watershed (Figure 2). The reasons for selecting this region as a case study for examining ecological drought are twofold. First, given the ecological values within the basin, including blue-ribbon trout streams and proximity to National Parks and wilderness areas, we expected that some drought plans would already address ecological impacts. Second, as discussed above, this region has been the focus of a pilot project for the National Drought Resilience Partnership. The Blackfoot watershed, which is outside the UMH boundaries, is included in our analysis because it was one of the earliest watersheds to adopt a "shared-sacrifice" model for addressing water shortages and is often referenced by water planners from the Montana Department of Natural Resources and Conservation (DNRC) as an excellent model for drought planning.

3. Methods

Of the seven watersheds in our study area, five have some form of drought plan [17,18,42–44] (Table 1). Currently, there are no drought plans for the Gallatin River or Madison River watersheds (The city of Bozeman, the largest town in the Gallatin Watershed, developed a Drought Management Plan in 2017. However, the scope of this city drought plan is different than a watershed drought plan and, therefore, was not included in this analysis. Efforts to update or create drought plans for all six UMH watersheds is currently underway as a part of the UMH-NDRP project). All six UMH watersheds and the Blackfoot watershed have restoration plans. We collected copies of all available plans and qualitatively coded and analyzed the plans.

Table 1. List of drought plans, year of publication, and associated watershed.

Plan Name	Year of Publication	Watershed
Jefferson River Watershed Committee (JRWC) Drought Management Plan [Jefferson Plan]	First published in 2000. Updated in 2007.	Jefferson and Boulder Rivers
Big Hole River Drought Management Plan and Plan Amendments (2002–2016) [Big Hole Plan]	First published in 2002. Amended through 2016.	Big Hole
Plan to Avoid Dewatering of the Ruby River Project [Ruby Plan]	1988	Ruby
Beaverhead Watershed Drought Resiliency Plan [Beaverhead Plan]	2016	Beaverhead and Red Rocks Rivers
Blackfoot Drought Response Plan [Blackfoot Plan]	Revised 2016	Blackfoot [1]

[1] The Blackfoot watershed is located outside the UMH Basin.

Our multi-disciplinary team (climatologists, ecologists, and social scientists) developed a drought plan coding scheme through an iterative process [45]. This coding scheme was informed by reviews of drought impact literature [46], the three pillars of drought management [27] and the framework for ecological drought [35]. We developed coding categories, co-coded sample plans, and compared inter-coder interpretations to arrive at common definitions and codes. The full coding scheme for each plan includes the planning background, overall plan approach, planning process, all drought impacts, climate change impacts, descriptions of vulnerability factors, indictors and monitoring, and plan actions. Each plan was coded independently by two team members, using NVivo 11 (QSR International). We assessed inter-coder reliability through team discussion and refinement of definitions and codes.

In this paper we focus on the degree to which each drought plan (1) identifies ecological impacts of drought (including mentions of impacts to fish, wildlife, other species, water quality, groundwater, wetlands, rivers, snowpack, wildfire, forests, grasslands, and ecosystem services), (2) identifies indicators of drought, and (3) states whether those indicators trigger drought mitigation or response actions. We summarized coded sections for each of these elements for further analysis, and recorded every mention of an ecological impact (Table 2) or drought indicator (Table 3) from each plan. Given that some impacts and indicators were thoroughly addressed while others were simply mentioned, we developed a scoring system for ecological impacts and indicators (Tables 2 and 3), following the criteria provided by Steinemann et al. [47]. Drought planning literature [26] emphasizes the need for specificity in the indicators and triggers portion of a drought plan. Plans should contain sufficient information about how indicators are measured, and the values used to trigger drought responses to assist current and future personnel in managing water resources during drought [48].

To help provide context for understanding the extent to which ecological impacts of drought were recognized by stakeholders within these watersheds, we also coded a set of Watershed Restoration Plans for the study area using the same coding scheme. In compliance with Section 319 of the federal Clean Water Act, these plans identify the causes and sources of water impairment within the watershed, identify indicators for monitoring water quality, and describe management activities that can help achieve watershed goals. This analysis allowed us to compare the list of impacts and indicators (Tables 2 and 3) from the drought plans to another list of ecological impacts and indicators, highlighting differences in the types of ecological drought impacts addressed. While this is a preliminary assessment, the comparison allows us to explore additional impacts and indicators used in other planning processes that could be applied to drought planning.

4. Results and Discussion

"When people say 'health of the river', what they really mean is the health of the fisheries in the river. [It's a] fairly narrow ecological view."

—local watershed coordinator, UMH

In our drought plan analysis, we found that some ecological impacts of drought were considered in all five drought plans (Table 2). Given the dominant focus in drought planning on agricultural and socio-economic impacts, the recognition of *non*-human impacts is an important finding. However, the analysis shows that the scope of ecological impacts mentioned within these watershed plans is generally limited to fish populations and fish habitat (Table 2) as monitored through two primary indicators: streamflows and water temperatures (Table 3). While this finding suggests a narrow view of ecological impacts, it is useful to show some examples of the way ecological impacts and indicators were discussed in the drought plans to provide more nuance.

4.1. Ecological Impacts

Some plans clearly demonstrate consideration of scientific information from the fields of ecology and fisheries biology. For example, the Blackfoot Plan discusses how drought affects fish populations, fish habitat, and briefly mentions native fish recovery and management. The plan provides detailed biological and ecological information, including the following description of the importance of riffles (rocky or shallow sections of a stream with rough or rippled water) for fish habitat:

Riffles are critical because they produce the chlorophyll (plant life) and forage (insects and small fish) that fuels the upper trophic levels (e.g., larger trout) of the ecosystem. In addition to basic river productivity, riffles provide spawning areas and habitat for juvenile trout and forage-fish alike. Entire communities—species ranging from midge to salmonfly, dace, sculpin and juvenile whitefish live in the cracks and crannies of cobbles that form the riffle. This forage base—the grocery list at the lower end of the food chain—sustains predatory species like trout as well as dependent wildlife in the upper food chain. When the wetted-width of the riffle narrows, river productivity rapidly declines and the forage base that sustains thriving trout fisheries is greatly diminished.

As the habitat base shrinks below minimal flows, it sets in motion a series of complex biological processes. These involve increased competition within fisheries communities for food and space; restricted movements between critical habitats (e.g., spawning sites and refugia); elevated mortality (at all trophic levels) as prey is concentrated; and cold-water communities become vulnerable to temperatures stressors depending on species and location. Juvenile fish are highly vulnerable to habitat loss and related stress and are the first to undergo population-level declines.

As flows decrease, water temperature increases. With elevated water temperature, metabolic rates increase and dissolved oxygen levels decline, pollutants concentrate and coldwater trout become more susceptible to pathogens like fungal infections and whirling disease. ([44], pp. 2–3)

While this description documents a broad understanding of the ecological impacts of drought across multiple species and processes in streams, the focus remains tied to the health of specific fisheries.

Similarly, the Big Hole Plan mentions broader ecological processes that can be affected by drought, such as changes in spring runoff conditions that are important for maintaining stream channels and moving sediment that would otherwise degrade riffles and other habitat features. This broader lens for understanding ecological impacts provides the potential for considering other interacting factors, such as land use near streams that can influence sediment movement. However, this plan, again, focuses on the relatively narrow objective of protecting to the health of fisheries and fish habitat, rather than considering how the stream ecosystem, and the watershed as a whole, may be at risk.

The Beaverhead Plan provides the most exhaustive coverage of ecological drought impacts including not only fish populations, but also wildlife, habitat, forest and range productivity, invasive weeds, forest fires, and ecosystem services. However, this exhaustive coverage of impacts did not translate into a high degree of specificity regarding how these impacts should be measured; nor did the ecological status of non-fisheries impacts trigger action (Tables 2 and 3). As a document designed to guide decision-making before, during, and after a drought, a drought plan should have specific indicators and actions. As Wilhite [26] notes:

Drought indicators and triggers are important for several reasons: to detect and monitor drought conditions; to determine the timing and level of drought responses; and to characterize and compare drought events. Operationally, they form the linchpin of a drought management plan, tying together levels of drought severity with drought responses. (p. 72)

Table 2. Ecological impacts mentioned in drought plans.

	Ecological Impacts Mentioned in Drought Plans	Blackfoot Plan	Big Hole Plan	Jefferson Plan	Ruby Plan	Beaverhead Plan
	Fish mortality or fish populations	3	3	3	1	3
	Fish habitat	3	2	-	1	1
	Water Quality	-	-	-	-	2
	Native Fish Recovery & Management	1	-	-	-	1
	Aquatic ecosystems	1	-	-	-	1
	Wildlife habitat	1	-	-	-	1
Ecological Impacts	Concentrated pollution	1	-	-	-	2
	Wildfire or forest fires	1	-	-	-	3
	Forest productivity	1	-	-	-	-
	Tree mortality	-	-	-	-	1
	Wildlife mortality or wildlife populations	1	-	-	-	1
	Non-ag, natural resource-based livelihoods	1	-	-	-	1
	Ecosystem services	-	-	-	-	1
	Weed pressure	-	-	-	-	1
	Range and forage productivity	1	-	-	-	1

Ecological impacts mentioned in drought plans were assigned a numerical value based on the level of detail provided in the plan and if it was associated with a specific indicator [- = Not mentioned in plan; 1 = Impact mentioned, but no details provided; 2 = Impact discussed in some detail, but no indicator mentioned; 3 = Impact discussed with an indicator for monitoring].

4.2. Ecosystem Services Impacts

As emphasized in the Ecological Drought Framework [35], ecological drought affects humans through its impacts on ecosystems services [49]. While this is a human-centric way of assessing the effects of drought, it provides a conceptual approach for broadening the range of ecological impacts considered. We found some examples of this approach in the drought plans we reviewed.

For example, the Beaverhead Plan acknowledges that the blue-ribbon trout fishery supports an important angling and tourism sector in the area. The plan notes that low stream flows and high water temperatures affect the fishery and outfitters, as well as "local hotels, restaurants, and other businesses"

([18], p. 14). The plan mentions that drought conditions can affect wildlife populations and hunting access (p. 36), forest lands and associated ecosystem services (p. 22), and causes "proliferation of noxious weeds" which costs time, money and, human resources to control (p. 62).

The Blackfoot Plan mentions impacts to natural resource-based livelihoods and the potential loss of income and jobs for those who "depend on water and other natural resources" ([44], p. 3). However, the discussion is limited to impacts on crop production, livestock, outfitters, and "other businesses that depend on visitors [who will seek] other areas for fishing and recreational opportunities" (p. 3). While a link could be made between drought and forest-dependent livelihoods (e.g., timber, wildlife hunting outfitters), this is not explicitly stated; only impacts to agricultural and river-based livelihoods are mentioned. Given that the water available for streams and agriculture is connected to snow abundance and related run-off from forested uplands [39], a lack of consideration of forest health could lead to further degradation of the system. However, the drought plans do not make this link.

4.3. Indicators and Triggers

As shown in Table 3, the most common indicators that trigger some type of drought mitigation or response action are streamflows (cubic feet per second [cfs] or gage height) and water temperature. The minimum streamflow requirements established in the Blackfoot Plan, the Big Hole Plan, and the Beaverhead Plan are based on the same "wetted stream perimeter" (or "wetted-riffle") method that is used to calculate instream flow rights for legal purposes (see Section 1.3 above). These plans also have water temperature thresholds that trigger a response designed to protect fish populations (e.g., angling restrictions or river closures). While streamflow and water temperature are the key indicators used to trigger mitigation or response actions, some plans mention other indicators that are used to assess drought conditions (Table 3).

The two-page Jefferson Plan is notable for its brevity, as well as its specificity. While it only uses streamflows and water temperatures as indicators, the plan clearly indicates where streamflow should be measured, i.e., "Twin Bridges Gaging Station (06026500)" ([42], p. 2). It also specifies what actions are triggered at the 600 cfs (e.g., voluntary conservation and angler awareness) and 280 cfs (e.g., possible fishing closure, voluntary reduction in irrigation and municipal water use, weekly meetings) stream flow levels. It notes that "the angling closure will remain in effect until flows reach or exceed 300 cfs for seven consecutive days at the Twin Bridges Gage" (p. 2). Additionally, "when maximum daily water temperature equals or exceed 73 degrees F (23 degrees C) for three consecutive days" the river will be closed to angling from 2pm to 12:00am (p. 2). These specific streamflow and temperature indicators are linked to specific actions and therefore score a three (Table 3). No other indicators are mentioned in this plan.

The Big Hole Plan is notable for the degree of specificity in describing how the wetted perimeter inflection points, which determine streamflow triggers, were calculated (see Plan Amendments, p. 6). Interestingly, the plan notes the complicated reality of enforcing ecologically-based minimum flow requirements, stating:

> The wetted stream perimeter (i.e., flow below which standing crops of fish decrease) of the upper Big Hole River is 60 cfs (DNRC 1992). While this flow may be reasonable to maintain in ample moisture years and should be the goal for flow preservation efforts, in most years it is not a realistic quantity. Fish population and flow data indicate 40 cfs is feasible to maintain while still sufficient to protect the Arctic grayling population. A minimum survival flow of 20 cfs will provide flows necessary to maintain a wetted channel, provide connectivity to thermal or flow refugia habitats, and ensure survival of the grayling population during brief, critical periods. ([43], p. 5)

The Big Hole Plan mentions snowpack and forecasted low stream levels and specifies the agencies that will provide this information (i.e., Montana Department of Natural Resources and Conservation [DNRC], U.S. Geological Survey [USGS], Montana Fish, Wildlife and Parks [MFWP], and Natural

Resources Conservation Service [NRCS]) (p. 2). However, the only indicators that are used to trigger actions are streamflows (cfs) and water temperature.

Table 3. Indicators mentioned in drought plans.

Indicators Mentioned in Drought Plans		Blackfoot Plan	Big Hole Plan	Jefferson Plan	Ruby Plan	Beaverhead Plan
Indicators	Streamflow (cfs or gage height)	3	3	3	3	3
	Water temperature	3	3	3	-	3
	Spring runoff	1	-	-	-	2
	Forecasted water supply, stream levels	-	2	-	1	3
	Other forecasted information	-	-	-	-	1
	Wetted-riffle or wetted stream perimeter	1	2	-	-	2
	Reservoir storage	-	-	-	1	3
	Snowpack or Snow Water Equivalent	1	2	-	1	2
	Precipitation	1	-	-	1	2
	CoCoRaHS rain gages	-	-	-	-	2
	Groundwater levels	-	-	-	-	2
	Air temperature	-	1	-	1	2
	Evapotranspiration	-	-	-	-	2
	Soil moisture	1	-	-	1	1
	Soil health	-	-	-	-	1
	Surface Water Supply Index (SWSI)	1	-	-	-	2
	Montana water supply index	-	-	-	-	2
	Biotic conditions	1	-	-	-	-
	Dissolved oxygen	1	-	-	-	-
	Forage production	-	-	-	-	1
	Basin scale wildfire risk indices	-	-	-	-	2
	Irrigation demand or ditch withdrawals	1	-	-	1	1
	US Drought Monitor (USDM)	-	-	-	-	3
	Palmer Drought Severity Index (PDSI)	-	-	-	-	1
	Drought Impact Reporter	-	-	-	-	2
	Drought Risk Atlas	-	-	-	-	1
	Gravity Recovery and Climate Experiment (GRACE)	-	-	-	-	1
	Normalized Vegetation Difference Index (NDVI)	-	-	-	-	1
	El Niño Southern Oscillation (ENSO) Outlook	-	-	-	-	2

Indicators were assigned a numerical value based on the degree of specificity used to describe it and if it was used to trigger a specific action [- = Not mentioned in plan; 1 = Indicator mentioned, but no description of how indicator is used; 2 = Indicator mentioned along with specific information about when, where, how and/or how often indicator is to be used and/or where the indicator information comes from; 3 = Indicator mentioned and used to trigger a specific action to mitigate or respond to drought].

This plan is unique in listing three different streamflow triggers for each river reach. The first trigger level is to "prepare for conservation" (e.g., present data, notify water users and anglers of low water conditions and encourage conservation, etc.), the second trigger level is to "conserve" (e.g., use Phone Tree to request that anglers limit their activities to cooler hours of the morning, contact media to inform public, etc.) and the third trigger level is for "river closure" (e.g., MFWP will close river sections, notify the public of closures, and encourage conservation, etc.).

In the upper and middle reaches of the Big Hole River, a fourth, and much earlier streamflow trigger is listed. These are reaches where Candidate Conservation Agreements with Assurances (CCAAs) are in place. The specified streamflow trigger (cfs) requires water users with CCAAs to implement their plans. Other water users are encouraged to implement conservation measures. On one particular river section (USGS Gage Number 06024580), a streamflow of 450 cfs triggers the implementation of CCAA plans. It is not until much lower flows of 170 cfs are reached that "preparation for conservation" is triggered. "Conservation" actions are triggered at 140 cfs (p. 6). In other words, the CCAA plans allow drought actions to be taken at a much earlier stage. Arguably, CCAA plans function as a mitigation strategy, rather than response strategy [50].

The Ruby Plan is the oldest plan, published in 1988. In contrast to the ecologically-based minimum streamflow levels in other plans, the Ruby Plan specifies that minimum streamflows levels were established to ensure that downstream senior irrigators receive their full water allotment. The primary indicator is gage height (i.e., streamflow). While water temperature is never mentioned in this plan, it does instruct the DNRC to "confer with the [Water User's] Association to discuss the snowpack, reservoir storage, streamflow, streamflow forecast, and soil moisture" (p. 7). The concept of "dewatering" suggests an early recognition that meteorological drought conditions combined with irrigation and water management decisions can affect downstream water availability.

The Blackfoot and Beaverhead Plans list a wider range of drought indicators (Table 3). However, the Blackfoot Plan states, "Stream flows are a primary indicator of drought conditions and can determine when specific actions under the Blackfoot Drought Plan will be implemented" ([44], p. 6). It adds that "Water temperature can also trigger drought response measures" (p. 6). The plan directs the Blackfoot Drought Committee to "examine other factors such as time of year, water demand, climatic conditions, weather projections and resource conditions" (p. 6). However, no further detail is provided about how, when or where these measurements will be taken. Interestingly, the plan states, "When all factors are considered, it is possible for stream flows and water temperatures to exceed trigger levels without the Drought Response being implemented" (p. 6).

The Beaverhead Plan, published in 2016, is the first drought plan to be completed as part of the National Drought Resilience Partnership project. While streamflows and water temperature are key indicators, this is the only plan that uses additional indicators—such as forecasted water supply, reservoir storage, and information from the U.S. Drought Monitor—to trigger actions. The plan provides website links and lists specific agencies or other resources where additional monitoring information can be found (these indicators received a score of two in Table 3). Additionally, the plan specifies indicators and information sources that are underdeveloped or underutilized in the watershed, including soil moisture, evapotranspiration, groundwater levels, streamflow gages for measuring reservoir inflow, and the Community Collaborative Rain, Hail and Snow (CoCoRaHS) network, which collects and reports precipitation measurements from their backyards of volunteer observers. The Beaverhead Plan calls for the development of a Drought Early Warning System website that would serve as a clearinghouse for information from all available climate and hydrology monitoring networks.

4.4. Other Resource Management Plans that Inform Drought Planning

The Beaverhead Plan describes several other resource management plans that identify drought-related vulnerabilities or have implications for drought resiliency, including, Western Governors' Associations Drought Forums, National Drought Forum Reports, State Water Plans, Bureau of Land Management (BLM) Resource Management Plans and Watershed Assessments, U.S. Forest Service (USFS) National Forest Plans, Watershed Restoration Plans, County-level Pre-Disaster Mitigation Plans (PDMS), and County Wildfire Protection Plans. As part of a preliminary analysis of potential synergies between planning documents, we conducted a brief review of the Watershed Restoration Plans for the seven watersheds in this study area (Figure 2), focusing again on ecological impacts and indicators. Relative to the drought plans, these plans significantly expand the range of ecological impacts that could be linked to drought resiliency, including interactions with water quality (eutrophication from nutrient inputs, sediment and pollutant loads), water temperature, whitebark pine populations, pest and pathogen outbreaks (e.g., mountain pine beetle outbreaks, white pine blister rust), invasive species, conifer encroachment, grassland productivity, soil erosion, wetlands, wildlife, and wildfire. These plans also expand the range of indicators that could be used in drought planning including measurements of nutrient or pollutant concentrations, macro-invertebrates, and streambank vegetation and shading, which can help to reduce water temperatures. Future research should examine a variety of resource management plans (BLM, USFS, DMPs, etc.) to better understand how, where, and by whom ecologically available water is being managed and how these plans can be used to improve watershed-level drought planning efforts.

5. Conclusions

In this paper, we have argued that in the western U.S., water laws and drought plans are used to prioritize and allocate scarce water resources. However, both have historically failed to recognize non-human water needs. Recent legislative efforts to establish instream flow rights at the state and federal levels are one mechanism for redressing the ecological harm caused by early water development.

Another mechanism is a new framework for drought planning that accounts for the non-human impacts of drought. Our analysis of five watershed drought plans in southwestern Montana found that while current plans do consider some of the ecological impacts of drought, it is generally through the narrow lens of impacts to fish populations and fish habitat as monitored through two primary indicators: water temperature and streamflow (Tables 2 and 3). The latter is typically based on the same ecological principles and methods that have commonly been used to develop minimum instream flow legislation.

While minimum instream flow legislation and drought plans that use minimum streamflows to trigger responses are vast improvements over the status quo, they have two major limitations. First, minimum flows often result in sub-optimal conditions for aquatic species and do not account for the dynamic nature of river ecosystems. Legislation that supports variable flow regimes, rather than minimum flows, can help address this issue. Second, the reliance on streamflow levels as the primary indicator of drought, encourages individuals and communities to focus on reactive drought strategies, rather than developing longer-term drought mitigation strategies. One exception is the use of CCAA plans which require drought actions (e.g., reducing irrigation diversions) to be taken at a much earlier stage, thus functioning more as a mitigation strategy than a response strategy.

The drought plans themselves have further limitations, including a lack of sufficiently specific indicators, a narrow consideration of ecological impacts that ignores other drought risks, and a missed opportunity to link to other resource planning processes. Given that livelihoods in this region are based on traditional agriculture, as well as angling, river recreation and tourism, it is not surprising that drought plans focus on the ecological value of fisheries. It may also be that fisheries impacts are more immediate and visible (e.g., floating dead fish) as compared to forest impacts and drought-induced tree mortality, which can be more complex, harder to see, and happen over longer time scales. Furthermore, existing and commonly used drought indicators (e.g., streamflows) are useful indicators for the health of fisheries. In contrast, indicators that may be more useful for monitoring impacts to rangeland and forest systems (e.g., soil moisture, Normalized Difference Vegetation Index [NDVI]) may be more expensive, less available, less familiar to drought planners, and/or less widely used.

Despite these criticisms, our findings suggest that the drought plans from southwestern Montana provide a starting point to account for the ecological impacts of drought, monitor for ecological impacts, and identify who manages ecologically available water. As the contrast between the 2016 Beaverhead Plan and earlier drought plans suggests, a holistic view of drought risk is likely to produce a more comprehensive approach to drought planning. The efforts of the NDRP have focused on building ongoing communication and strong relationships between local watershed drought planners and other resource managers who develop and use plans such as Watershed Restoration Plans, BLM Watershed Assessments, and USFS Forest Plans. As these efforts continue and synergies emerge, the hope is the new drought plans will reflect the full range of ecological impacts, include the suite of indicators available for measuring drought conditions, and focus on mitigation strategies that increase the resiliency of both human and natural systems. Our analysis of existing drought plans in the UMH emphasizes the importance of integrating these three elements. Furthermore, it is critical to understand that watershed communities are composed of humans, as well as the non-human entities that co-habit this world; both are affected when conditions are hot and dry. Watershed planning efforts in the UMH are well on their way to incorporating a diversity of drought impacts, but would benefit from a deeper consideration of the health and functioning of a range of ecological processes that take place within and around streams, waterways, and riparian areas.

Our findings contribute to recent work by Mount et al. [30] on ecological drought in California which calls for the development of "watershed-level plans that set ecosystem priorities and identify trade-offs" (p. 3) and echoes their conclusion that better integration of various planning processes will likely improve drought preparedness. These findings are applicable to arid and drought-prone regions across the globe.

Acknowledgments: This research was supported in part by the U.S. Geological Survey (USGS) under Grant/Cooperative Agreement G15AC00277 and in part by the Science for Nature and People Partnership

(SNAPP)—a partnership of The Nature Conservancy, the Wildlife Conservation Society, and the National Center for Ecological Analysis and Synthesis (NCEAS) at the University of California, Santa Barbara. Any use of trade, firm, or product names is for descriptive purposes only and does not imply endorsement by the U.S. Government. A.E.C. and J.M. received publication credit from MDPI which covered a portion of the publication costs.

Author Contributions: J.M., D.J.B., A.E.C., T.H., K.R.H. and M.J.H. conceived and designed the research project; J.M., D.J.B., A.E.C., T.H., K.R.H., M.J.H., T.J., M.P. and E.W. developed the coding scheme, coded the plans, and analyzed the data; E.W. created the map in Figure 2. J.M. and M.P. created the impacts and indicator tables. T.J. managed the citation database. J.M. led the writing of the paper and wrote the results, discussion, and conclusion sections; M.J.H. and D.J.B. wrote the section on the history of drought planning, A.E.C. wrote the section on water law and case study background, N.B. wrote the section on instream flow policies; T.H. wrote the methods section; all authors contributed significantly to the writing and editing of the paper.

Conflicts of Interest: The authors declare no conflict of interest. Several of the authors are members of the SNAPP working group on Ecological Drought. Support for meetings of the SNAPP working group that led to this publication was provided by the USGS. A.E.C. and N.B. are USGS Scientists. The USGS grant included some salary support for K.R.H.

References

1. Thompson, B.H., Jr.; Leshy, J.D.; Abrams, R.H. *Legal Control of Water Resources: Cases and Materials*; West Pub. Co./Thomson Reuters: St. Paul, MN, USA, 2013.
2. Fanning, W.; Sime, C.; Mudd, M.B.; Williams, M. Water Rights in Montana: How our legal system works today, how Montana compares to other states, and ideas for Montana's future. In *Report for the Montana Supreme Court*; University of Montana School of Law Land Use & Natural Resources Clinic: Missoula, MT, USA, 2017.
3. *Wild and Scenic Rivers Act (PL 90-542)*; 16 U.S.C. 1271 et seq; Legal Information Institute: Ithaca, NY, USA, 1968.
4. *Clean Water Act*; 33 U.S.C §§ 1251 et seq; Legal Information Institute: Ithaca, NY, USA, 1977.
5. *Murphy Water Rights (89-901 RCM)*; Section 2; Montana River Action: Bozeman, MT, USA, 1969.
6. *Montana Water Use Act of 1973*; Title 85; Chapter 2; Montana Code Annotated (MCA): 1973. Available online: http://leg.mt.gov/bills/mca_toc/85_2_3.htm (accessed on 19 February 2018).
7. Stalnaker, C.; Lamb, B.L.; Henriksen, J.; Bovee, K.; Bartholow, J. *The Instream Flow Incremental Methodology (Biological Report 29: March 1995)*; U.S. Department of the Interior National Biological Service: Washington, DC, USA, 1995.
8. Gillian, D.M.; Brown, T.C. *Instream Flow Protection: Seeking a Balance in Western Water Use*; Island Press: Washington, DC, USA, 2013.
9. Council, N.R. *The Science of Instream Flows: A Review of the Texas Instream Flow Program*; The National Academies Press: Washington, DC, USA, 2005; ISBN 978-0-309-09566-2.
10. McKinney, M.J. Instream Flow Policy in Montana: A History and Blueprint for the Future. *Public Law Rev.* **1990**, *11*, 81–134.
11. *Application for Permit or Change in Appropriation Right*; 85-2-302; Montana Code Annotated (MCA): 2015. Available online: http://leg.mt.gov/bills/mca/85/2/85-2-302.htm (accessed on 19 February 2018).
12. Zellmer, S. Legal tools for instream flow protection. In *Integrated Approaches to Riverine Stewardship: Case Studies, Science, Law, People, and Policy*; Instream Flow Council: Cheyenne, WY, USA, 2008.
13. United States Fish & Wildlife Service. Candidate Conservation Agreements. Available online: https://www.fws.gov/endangered/esa-library/pdf/CCAs.pdf (accessed on 19 February 2018).
14. Montana Department of Fish, Wildlife, and Parks; U.S. Fish and Wildlife Service; Montana Department of Natural Resources and Conservation; U.S. Natural Resources Conservation Service. *Candidate Conservation Agreement with Assurances for Fluvial Artic Grayling in the Upper Big Hole River: FWS Tracking #TE104415-0 (March 30th, 2006)*; MT Fish Wildlife and Parks: Helena, MT, USA, 2006.
15. Anderson, M.B.; Ward, L.; McEvoy, J.; Gilbertz, S.J.; Hall, D.M. Developing the water commons? The (post)political condition and the politics of "shared giving" in Montana. *Geoforum* **2016**, *74*, 147–157. [CrossRef]
16. Anderson, M.B.; Ward, L.C.; Gilbertz, S.J.; McEvoy, J.; Hall, D.M. Prior appropriation and water planning reform in Montana's Yellowstone River Basin: Path dependency or boundary object? *J. Environ. Policy Plan.* **2017**, 1–16. [CrossRef]

17. Ruby Valley Conservation District Office. *Plan to Avoid Dewatering of the Ruby River Project*; Ruby Valley Conservation District Office: Sheridan, MT, USA, 1988.
18. Carparelli, C. Beaverhead Watershed Drought Resiliency Plan. 2016. Available online: http://www.beaverheadwatershed.org/wp-content/uploads/2017/08/Beaverhead-Drought-Resiliency-Plan-2016.pdf (accessed on 31 January 2018).
19. Wishart, D.J. *The Last Days of the Rainbelt*; University of Nebraska Press: Lincoln, NE, USA, 2013.
20. NOAA/NCEI-2 Billion-Dollar Weather and Climate Disasters: Overview. Available online: https://www.ncdc.noaa.gov/billions/ (accessed on 19 February 2018).
21. Wilhite, D.; Svoboda, M.; Hayes, M. Understanding the complex impacts of drought: A key to enhancing drought mitigation and preparedness. *Water Resour. Manag.* **2007**, *21*, 763–774. [CrossRef]
22. Hayes, M.; Wilhite, D.; Svoboda, M.; Trmka, M. Investigating the Connections between Climate Change, Drought, and Agricultural Production. In *Handbook on Climate Change and Agriculture*; Dinar, R.M.A., Ed.; Edward Elger Publishing: Cheltenham, UK, 2011; pp. 73–86.
23. Smith, K.H.; Svoboda, M.; Hayes, M.; Reges, H.; Doesken, N.; Lackstrom, K.; How, K.; Brennan, A. Local Observers Fill in the Details on Drought Impact Reporter Maps. *Bull. Am. Meteorol. Soc.* **2014**, *95*, 1659–1662. [CrossRef]
24. Cook, E.R.; Woodhouse, C.A.; Eakin, C.M.; Meko, D.M.; Stahle, D.W. Long-Term Aridity Changes in the Western United States. *Science* **2004**, *306*, 1015–1018. [CrossRef] [PubMed]
25. Overpeck, J.T. Climate science: The challenge of hot drought. *Nature* **2013**, *503*, 350–351. [CrossRef] [PubMed]
26. Wilhite, D.A.; Hayes, M.J.; Knutson, C.L. Drought Preparedness Planning: Building Institutional Capacity. In *Drought and Water Crises: Science, Technology, and Management Issues*; Wilhite, D.A., Ed.; Taylor & Francis/CRC Press: Boca Raton, FL, USA, 2005.
27. Wilhite, D.A. *National Drought Management Policy Guidelines: A Template for Action. Integrated Drought Management Programme (IDMP) Tools and Guidelines Series 1*; Global Water Partnership: Stockholm, Sweden, 2014.
28. Wilhite, D.A.; Buchanan-Smith, M. Drought as a natural hazard: Understanding the natural and social context. In *Drought and Water Crises: Science, Technology, and Management Issues*; Wilhite, D.A., Ed.; Taylor & Francis/CRC Press: Boca Raton, FL, USA, 2005.
29. Gartrell, G.; Mount, J.; Hanek, E.; Gray, B. *Approach to Accounting for Environmental Water: Ingishts from the Sacramento-San Joaquin Delta*; Public Policy Institute of California: San Francisco, CA, USA, 2017.
30. Mount, J.; Gray, B.; Chappelle, C.; Gartrell, G.; Grantham, T.; Seavy, N.; Szeptycki, L.; Thompson, B.B. *Managing California's Freshwater Ecosystems: Lessons from the 2012–2016 Drought*; Public Policy Institute of California: San Francisco, CA, USA, 2017.
31. Mount, J.; Gray, B.; Chappelle, C.; Doolan, J.; Grantham, T.; Seavy, N. *Managing Water for the Environment During Drought: Lessons from Victoria Australia (Report)*; Public Policy Institute of California: San Francisco, CA, USA, 2016.
32. Van Dijk, A.I.J.M.; Beck, H.E.; Crosbie, R.S.; de Jeu, R.A.M.; Liu, Y.Y.; Podger, G.M.; Timbal, B.; Viney, N.R. The Millennium Drought in southeast Australia (2001–2009): Natural and human causes and implications for water resources, ecosystems, economy, and society. *Water Resour. Res.* **2013**, *49*, 1040–1057. [CrossRef]
33. United Nations Environment Programme. *Devastating Drought in Kenya: Environmental Impacts and Responses*; UNEP: Nairobi, Kenya, 2000.
34. Islam, S.N.; Tuli, S.M. Drought Impacts on Urbanization in. In *Handbook of Drought and Water Scarcity: Environmental Impacts and Analysis of Drought and Water Scarcity*; CRC Press: Boca Raton, FL, USA, 2017; pp. 17–44. ISBN 9781498731041.
35. Crausbay, S.D.; Ramirez, A.R.; Carter, S.L.; Cross, M.S.; Hall, K.R.; Bathke, D.J.; Betancourt, J.L.; Colt, S.; Cravens, A.E.; Dalton, M.S.; et al. Defining ecological drought for the 21st century. *Bull. Am. Meteorol. Soc.* **2017**, 2543–2550. [CrossRef]
36. Hayes, M.J.; Svoboda, M.D.; Wardlow, B.D.; Anderson, M.C.; Kogan, F. Drought Monitoring: Historical and Current Perspectives. In *Remote Sensing of Drought: Innovative Monitoring Approaches*; Wardlow, B.D., Anderson, M.C., Verdin, J.P., Eds.; CRC Press/Taylor & Francis: Boca Raton, FL, USA, 2012.
37. Van Loon, A.F.; Gleeson, T.; Clark, J.; Van Dijk, A.I.J.M.; Stahl, K.; Hannaford, J.; Di Baldassarre, G.; Teuling, A.J.; Tallaksen, L.M.; Uijlenhoet, R.; et al. Drought in the Anthropocene. *Nat. Geosci.* **2016**, *9*, 89–91. [CrossRef]

38. Dunham, J.; Angermeier, P.; Crausbay, S.; Cravens, A.E.; Gosnell, H.; McEvoy, J.; Moritz, M.; Raheem, N.; Sanford, T. Rivers and Social-Ecological Systems: Time to Integrate Human Dimensions into Riverscape Ecology and Management. *WIREs Water*, in press.

39. Montana Department of Natural Resources and Conservation. *Montana State Water Plan: A Watershed Approach to the 2015 Montana State Water Plan*; Montana Department of Natural Resources and Conservation: Helena, MT, USA, 2015.

40. Montana Drought Demonstration Partners. *A Workplan for Drought Resilience in the Missouri Headwaters Basin: A National Demonstration Project*; 2015. Available online: http://dnrc.mt.gov/divisions/water/management/docs/surface-water-studies/workplan_drought_resilience_missouri_headwaters.pdf (accessed on 19 February 2018).

41. The Montana Watershed Coordination Council. Available online: Mtwatersheds.org (accessed on 19 February 2018).

42. Jefferson River Watershed Council. JRWC Drought Management Plan, 2012. Available online: http://jeffersonriverwc.com/fish/uploads/2016/06/JRWC_Drought_Mgt_Plan_2012.pdf (accessed on 31 January 2018).

43. The Big Hole Watershed Committee. *Big Hole River Drought Management Plan*; The Big Hole Watershed Committee: Whitehall, MT, USA, 2016. Available online: http://fwp.mt.gov/fwpDoc.html?id=68835 (accessed on 31 January 2018).

44. Blackfoot Challenge. *Blackfoot Drought Response Plan*; Blackfoot Challenge: Ovando, MT, USA, 2016. Available online: http://www.blackfootchallenge.org/Clone/wp-content/uploads/2012/06/Blackfoot-Drought-Response-Plan.pdf (accessed on 31 January 2018).

45. Berkowitz, S. Analyzing Qualitative Data. In *User-Friendly Handbook for Mixed Method Evaluations*; Frechtling, J., Sharp, L., Eds.; National Science Foundation, Division of Research, Evaluation and Communication: Arlington, VA, USA, 1997.

46. Vose, J.M.; Clark, J.S.; Luce, C.H.; Patel-Weynan, T. *Executive Summary. Effects of Drought on Forests and Rangelands in the United States: A Comprehensive Science Synthesis*; Gen. Tech. Rep. WO-93a; United States Department of Agriculture: Washington, DC, USA, 2015; p. 10.

47. Steinemann, A.C.; Hayes, M.J.; Cavalcanti, L.F.N. Drought indicators and triggers. In *Drought and Water Crises: Science, Technology, and Management Issues*; Taylor and Francis: Boca Raton, FL, USA, 2005; pp. 71–92.

48. Wilhite, D.A.; Glantz, M.H. Planning for drought: A methodology. In *Drought Assessment, Management, and Planning: Theory and Case Studies*; Kluwer Academic Publishers: Dordrecht, The Netherlands, 1993; pp. 87–108.

49. Raheem, N.; Cross, M.S.; Bathke, D.J.; Cravens, A.E.; Crausbay, S.; Ramirez, A.; McEvoy, J.; Carter, S.; Rubenstein, M.; Schwend, A.; et al. Planning for Ecological Drought Effects on Ecosystem Services: An Example from the Upper Missouri Headwaters Basin in Montana, USA. *WIREs Water*. under review.

50. Rossi, G. Drought Mitigation Measures: A Comprehensive Framework. *Adv. Nat. Technol. Hazards Res.* **2000**, *14*, 233–246.

resources

MDPI

Review

Rights of Nature: Rivers That Can Stand in Court

Lidia Cano Pecharroman

AC4, Earth Institute, Columbia University, New York, NY 10027, USA;
cano.lidia@columbia.edu; Tel.: +1-917-215-0178

Received: 7 December 2017; Accepted: 1 February 2018; Published: 14 February 2018

Abstract: An increasing number of court rulings and legislation worldwide are recognizing rights of nature to be protected and preserved. Recognizing these rights also entails the recognition that nature has the right to stand in court and to be represented for its defense. This is still an incipient field and every step taken in this direction constitutes a precedent from which to learn and on which to base new rulings and legislation initiatives. Within this doctrine, rivers seem to be on the spotlight and court rulings on the rights of rivers are the ones setting precedent. These cases have taken place in New Zealand, Ecuador, India, and Colombia. This review looks into what all these rulings and legislation worldwide say about the rights of nature and what legal and systemic considerations should be taken into account as the recognition of the rights of nature moves forward.

Keywords: rights; nature; rivers; Yamuna; Ganges; Atrato; Vilcabamba; Whanganui; jurisprudence; earth centric; deep ecology; law

1. Introduction: The Rights of Nature—What, Why and How?

Recognizing that nature has legal rights and accepting these rights as part of our legal systems require not only the introduction of new laws observing these rights, but also a shift in paradigm for them to be fit in a contemporaneous legal puzzle. Referring to a "shift" in paradigm and not the "introduction" of a newly created one is intentional, as the recognition of rights to nature has been already part of customary law for many indigenous populations around the world for centuries. These principles, however, have not been embedded in the development of modern environmental laws, grounded on an anthropocentric paradigm. This paradigm, has proven to be erroneous, as humans are irreversibly damaging the natural structures they depend on for survival despite the existence of environmental laws. The Paris Agreement and the planetary efforts that have been recently undertaken to move away from this approach and to develop in a sustainable manner point towards the possibility of a shift towards an earth-centered paradigm, where humans are part of the planetary system and aim at living in harmony with it. As mentioned before, the idea of making the rights of nature part of the way humans conceive their reality and rule their community is not new. Neither is the attempt to introduce this concept in modern legal systems. This essay reviews a suite of philosophical and practical efforts to recognize the rights of nature over the past 50 years, focusing on several recent cases involving rivers around the world.

The first scholar to raise the question of whether nature should be recognized the right to stand in court was Professor Christopher Stone, a professor from the University of Southern California who in 1972 wrote his famous essay: "Should Trees Have Standing? Toward Legal Rights for Natural Objects" [1]. To provide some background on the premises of Stone's essay, the Sierra Club had recently tried to sue Walt Disney Enterprises to prevent the construction of a ski resort in Mineral King Valley (in the Sierra Nevada Mountains). The US Court of Appeals in California responded, pointing out that the Sierra Club itself had not been injured by the project and as a result it did not possess the right to stand in court to bring a lawsuit against the corporation [2]. In response to the court's decision, Stone's essay expounded a robust set of reasons why the legal system should recognize nature's right to stand in court: to prevent cases like this in which neither environmental groups nor nature itself could be defended against damages in court. He introduces the topic by admitting that such a proposal might seem "frightening or laughable" as have any previous attempts to confer rights onto other entities in the past. A little more than a century ago, he argues, the majority of Americans were not outraged when a court would argue that Blacks were denied the rights of citizenship because they constituted "a subordinate and inferior class of beings, who had been subjugated by the dominant race" [3] or when another responded to a woman's willingness to become a lawyer that "the law of nature destines and qualifies the female sex for the bearing and nurture of the children [...] and all life-long callings of women, inconsistent with these radical and sacred duties of their sex, as is the profession of the law, are departures from the order of nature" [3]. These statements would sound outrageous if employed nowadays, but there was a time when they constituted legal jurisprudence. Women, slaves, or African Americans, were once rightless but as Stone reaffirms it is not "until the rightless thing receives its rights, [...that we can] see it as anything but a thing for the use of "us"—those who are holding rights at the time." [1] A similar argument is drawn by Steven Wise when providing reasons for why animals should also bear rights. He discusses that women, children, and slaves were once considered "legal things" and hence did not have the right to stand in court [4]. However, as our legislation evolves, in a similar fashion, animals with practical autonomy should too have rights. Furthermore, philosopher Peter Singer adds to this argument, arguing that, when recognizing more rights to humans than to animals in similar situations, we are being biased in favor of our own species (a phenomenon that he calls "speciesism") [5]. In the same manner, it is not until nature is recognized as holding certain rights that we will realize that nature is deserving of a chance to speak for itself.

Anticipating the counterargument that nature cannot stand in court as it is not a being, Stone and other scholars have a straightforward answer. Corporations, municipalities, and other entities have rights and can stand in court in the current legal system, so in the same way nature could be recognized certain rights and be represented in court. For those who counter that argument by noting that at least companies and governments can bear duties if asked for it in court, authors like Leimbacher reject the idea that there must exist a link between legal subjectivity and the ability to bear duties [6]. For instance, children have the right to be represented in court but do not bear any responsibility. Sitter goes further to clarify that legal subjectivity does not have to be linked to the idea to safeguard rights personally, as it is the case with nature.

Others may argue that there is not enough ground to justify the recognition of the rights of nature and that our legal systems were not designed for nature to be a holder of rights. But some authors have addressed these questions. Nedelsky, for instance, calls for the need for a new system to define rights. Rights should be defined in terms of relationships rather than the individuals that withhold said rights. As opposed to the idea that rights are a set of timeless and immutable values that already exist, instead rights constitute an intricate system of relationships that keep evolving. For instance, until recently, "great restrictions on the legal rights and opportunities for women were [in fact] believed to be consistent with a basic commitment to equality" [7]. The evolution of these relationships allows us to recognize new rights. These relationships do not only exist amongst humans but are also established when we interact with nature. As a result, we are able to recognize its rights as well as our duties towards it. Stone goes beyond this to define what it would mean for nature to be a "holder of legal rights" and establishes three requirements. First, the holder can institute legal actions on its behalf. Second, a court must take injury to it into account when granting redress. Third, this redress has to be to the benefit of it [1]. These definitions are nontrivial. They shed light on the basis for the rights of nature and the extension of these rights if applied.

Beyond the philosophical grounds for the rights of nature to stand in court, it is important to clarify what it means to hold these rights in legal terms. When venturing into such a debate, it is important to clarify the distinction between the legal institution of "personhood" and the implications of holding "locus standi." In the continental law tradition, holding legal personhood means holding a set of rights and duties. As individuals, every (physical) person holds legal personhood and has the rights and duties determined by the law. Similarly, a legal person, formed by a set of people or goods, holds legal personhood and is also given rights and duties by the law. Exercising these rights and duties would be the next step, but not every person holds legal competence to do so on its own. Some persons both physical and legal are deemed incapable of exercising such rights on their own. This could be due to their nature, as in the case of a company. It could also be a temporary situation, such as the case of children who will be able to fully exercise their rights and duties on their own once they become adults [8]. In both these cases, the full exercise of their rights takes place via a legal representative. Once a matter is brought to court, the parties interested in participating in the litigation process must demonstrate that the action challenged or the law put into question is affecting them directly or has some reasonable connection to their situation. The requirements and regulations in this regard vary across legal systems, but the overall idea is the same. To stand in court you need to hold legal personhood, be capable of exercising your rights, or have a legal representative otherwise, and hold locus standi.

So, where does nature fit into these definitions? As of today, there is no clear answer to this question. This discussion is currently being held worldwide in debates that courts, legislators, and international organizations are trying to settle. Many authors have called for a more flexible and inclusive theory for defining legal personhood that includes animals and nature. Some propose that the concept of legal personhood moves from a binary system [9] (holding or not legal personhood) to a system in which somebody or something can hold legal personhood to a certain degree [10]. In Korsgaard's words "for various, different, kinds of reasons, it seems inappropriate to categorize a fetus, a non-human animal, the environment, or an object of great beauty, as a person, but neither does it seem right to say of such things that they are to be valued only as means" [9]. Some authors have directly attributed the lack of flexibility in the way we define legal personhood to the fact that this concept "has long been associated with humanity, and even the paradigm of [...] corporations relies upon analogizing to humanity" [11]. Moving away from this paradigm would mean to "divorce the capacities-focused definition of legal personhood from the species-based definition of humanity" [11].

Finally, besides more theoretical approaches to whether rights of nature should and can be recognized by our legal systems, there is a more utilitarian approach to why these rights are needed to champion a protection of nature that has not been accomplished by the current legal system. Leimbacher argues that a relationship of dominance between humans and nature can never put a stop to further damaging nature [6]. The rights of nature from a utilitarian perspective are seen as a way to guarantee human's right to exist, as protecting nature, on which we depend, is a way to protect human interests. Leimbacher and other authors have linked the need to recognize the rights as a way to protect human rights to existence. Stone also explains that, because the health and well-being of humankind depend on the health of the environment, these goals will often be so mutually supportive that one can avoid deciding whether our rationale is to advance "us" or a new "us" that includes the environment. Other scholars such as Colon Ríos base the need to insert the rights of nature into our legal system on the theory of constituent power. Rights are the main tool to preserve conditions that are essential to the future exercise of rights by future generations. The environment will enable the development of future generations. Considering nature as a means for life has been a way to protect nature, protecting the right to a "healthful environment" and the human right to health [12]. In other words, we shall respect and protect the environment in which the next generations will flourish.

The legal and philosophical debates to define the rights of nature, how they should be applied, and the role they play in our current legal system is still open. However, some countries and international organizations are already including them in their legislations and debates at a national and international level.

2. The Rights of Nature at a Local, National, and International Level

Worldwide there is an increasing body of legislation, jurisprudence, and political declarations challenging the current paradigm and recognizing certain rights to nature. The process is still incipient and therefore it is important to examine the work that has already been done: first, to have a sense of where things stand and how far the rights of nature have permeated our society, second, to have a sense of how much work needs to be done in this realm, and third, to learn from the precedents when trying to include the rights of nature in new legislation. It is important to look both at local and national legislation, as well as the efforts carried out by the international community and the precedents set by indigenous populations that have embedded their entire legal systems in this paradigm for centuries.

The first place ever to recognize nature's right in an ordinance to protect the "citizens and environment [. . .] general welfare" was the Tamaqua Borough (PA, USA) [13]. The ordinance sought to ban the dumping of toxic sewage sludge in the community as a violation of the rights of nature [12]. Indirectly, the regulation hints towards the recognition of nature rights by recognizing that "everyone has the right to a healthy, protected, and balanced environment" and the exercise of these rights must be granted to individuals, collectives, and to "other living things so they may develop in a normal and permanent way." Only two years later, the country that pioneered the inclusion of the rights of nature as a constitutional right was Ecuador. In 2008, they approved a new constitution with which the country aimed at "building a new way of coexistence amongst citizens, in diversity and harmony with nature" [14]. The constitution has a chapter [14] exclusively dedicated to the rights of nature. The text states that nature has the right to be respected, and that its existence and the maintenance and regeneration of its life cycles, structure, and evolving processes must be allowed for. Furthermore, it gives any person the right to ask public authorities to respect its rights. Moreover, the constitution states that the state will apply "precautionary" and "restrictive" measures to any activity that may lead to the extinction of a species, the destruction of the ecosystems, or the permanent alteration of natural cycles. These rights have already been used as a reference to interpret legislation in other matters. When a plaintiff asked the constitutional court to rule the Organic Law of Special Regime for the Conservation and Sustainable Development of the Province of Galapagos unconstitutional for prioritizing nature protection over Ecuadorians rights to internal migration, the Constitutional Court used the rights of nature to argue otherwise. The court ruled against the plaintiff highlighting, among other things, that the constitutional articles on the rights of nature are the basis for every person to demand from authority the fulfillment of said rights, as well as the encouragement and promotion of respect for the ecosystems [15], over internal migration rights in this case. The constitution is clear when it comes to proclaiming nature as an entity that holds rights (the right to be respected, to be taken care of, etc.). However, as constitutional principles remain broad, it is unclear how these rights would be exercised, and whether or when nature would hold locus standi to defend these rights. Only a year after Ecuador made such constitutional changes, Bolivia's constitution also included rights of nature as part of its new text: Section I: Environmental Rights (protection of environment) includes the following: "Article 33: Everyone has the right to a healthy, protected, and balanced environment. The exercise of this right must be granted to individuals and collectives of present and future generations, as well as to other living things, so they may develop in a normal and permanent way. Article 34: Any person, in his own right or on behalf of a collective, is authorized to take legal action in defense of environmental rights, without prejudice to the obligation of public institutions to act on their own in the face of attacks on the environment" [16]. In 2010, Bolivia approved the Law of the Rights of Mother Earth [14] and the Framework Law of Mother Earth and the Integral Development of Living Well (Ley 300 (2012), Ley Marco de la Madre Tierra y el Desarrollo Integral para Vivir Bien). The 2010 law created an Ombudsman for the Rights of Mother Earth (Defensoría de la Madre Tierra), an institution in charge of safeguarding the rights established in the law [17]. Moreover, the framework establishes a set of institutions that will take action if the rights of nature are violated. This entitles citizens to take legal action, as long as the government is not doing so first (Article 39 of Law 300 (2012), Framework Law of the Mother Earth and for the Integral Development for Living Well). Paradoxically, this law, even though it is dedicated to the establishment of the rights of nature, also states that people have rights to the exploitation of natural resources as long as permission is granted by the government. Unless these concepts are further developed legally, this approach continues to perpetuate the conventional legal approach where the exploitation of nature is allowed as long as it respects environmental regulations in place. Indigenous groups in the country have already expressed their disappointment at the lack of substance of the law [17]. The indigenous leaders of the Isiboro Sécure Indigenous Territory and National Park (Tipnis) recently presented a complaint before the International Court for the Rights of Nature against the Government of Evo Morales for the depredation of its territory [18].

In the international arena, this topic is gathering an increasing amount of attention. The efforts have focused on raising awareness on the topic and creating political resolutions to support the concept of the rights of nature, rather than getting closer to an actual definition of its legal meaning and implications. Nevertheless, these political declarations are an important first step towards the development of legislation at a national and local level. The General Assembly (GA) adopted its first resolution on Harmony with Nature in 2010, encouraging Member States to make use of the designated "Mother Earth Day" [19] and including Harmony with Nature as part of the Assembly's agenda on Sustainable Development. The Assembly, however, did not mention the concept of "rights of nature" per se until the following resolution in 2011 [20]; in the resolution, the GA, "Noting the first Peoples' World Conference on Climate Change and the Rights of Mother Earth," created an interactive dialogue to share national experiences on how to approach and measure sustainable development in harmony with nature [20]. These dialogues have been held annually on Mother Earth's Day (22 April) as a forum where scholars, civil society, and diplomatic representatives have discussed the concept of Harmony with Nature and the Rights of Nature. This constant activity and the interest of some countries to further deepen the concept of the rights of nature led the GA in its 2016 [20] resolution on Harmony with Nature to note that "some countries recognize the rights of nature in the context of the promotion of sustainable development, and express the conviction that, in order to achieve a just balance among the economic, social, and environmental needs of present and future generations, it is necessary to promote harmony with nature."

The rest of the international community has slowly followed through with this concept in the last three years. In 2014, the G77 [21] signed the letter "For a New World Order for Living Well," in which these countries recognized that "Earth and its ecosystems are our home." In the letter, "some countries recognize the rights of nature in the context of the promotion of sustainable development" and called for a holistic approach to development that may include the recognition of these rights to restore the integrity of the Earth's ecosystems. The International Union for Conservation of Nature (IUCN), for instance, seeks to move towards the recognition of the rights of nature. The IUCN Programme for 2017–2020 (approved in 2016) states that it "aims to secure the rights of nature and the vulnerable parts of society through strengthening governance and the rights-based approach to conservation." Doing so sets as part of its Target 6 to "raise awareness of the rights of nature, and the cultural and spiritual values of nature," and emphasizes the need to "include urban populations and youth in understanding nature's intrinsic and intangible values" and "to advance rights regimes related to the rights of nature" [22].

The existence of this legal and political precedence is sparking actions for the recognition of the rights of nature in other parts of the world, both on national and local levels. For instance, Mexico City recently approved its own constitutional text that recognizes citizens' rights to a healthy environment, determines that is the government's duty to protect nature, and establishes that "a secondary law will be issued to recognize and regulate the broader protection of the rights of nature conformed by all its ecosystems and species as a collective entity subject to rights" (See Article 13 of the Mexico City Constitution.). In Argentina, a senator has proposed a bill to recognize the rights of nature in the national legislation [23]. Similarly, in Brazil, a draft amendment has been presented whereby Sao Paolo "will promote the development of environmental policies, considering that members of nature have inherent rights to life and maintenance of their ecosystem processes" [24]. In Europe, a group of lawyers, environmentalists, and academics are organizing a so-called "European Citizens Initiatives" [25] to propose the adoption of a law that "recognizes the right of nature to exist, renew, and maintain its vital cycles" [26].

Finally, it would not be fair to talk about the legal doctrine of the rights of nature without mentioning the traditions and knowledge on this issue that indigenous populations have passed onto generations. The paradigm that embraces and understands nature as a being with rights has been part of many indigenous populations' worldviews for hundreds of years. Their interdependent relationship with nature has resulted in non-anthropocentric social systems in which human's harmonious relation with nature has been always the desirable outcome [27].

This view of the world is embodied by the Sumak Kawsay or living well, a prevalent way of life across Latin American indigenous populations. These communities see nature as the Pacha Mama, a being formed by the harmonious interactions of the beings and natural systems on Earth (Article 5.1. Law 300 (2012), Ley Marco de la Madre Tierra, Plurinational State of Bolivia). For some of them, Pacha Mama is considered a deity, the mother of humanity that humans should respect and take care of to continue living in harmony. The understanding of nature as the Pacha Mama is not only the idea of it as a deity, it is rather a philosophy of life [28]. It is a way of living in harmony with nature, co-existing with it, caring for it, and allowing for its regeneration to provide for the upcoming generations [29]. This conception of reality is part of the social fabric [30] and as a result it permeates all the norms that govern the way these communities live [31]. These deep ecology conceptualizations of life can be found in every corner of the world. For instance, the Maori in New Zealand express this reciprocal and equal relation between humans and nature with their saying "Ko au te awa, ko te awa ko au," which translates to "I am the river and the river is me" [32].

In the African continent, this philosophy of reciprocity and respect for nature is also ingrained in the traditions of indigenous populations that conceive nature as sacred and conceive the use of its resources as long as nature can regenerate. These societies contemplate within their norms the denomination of a custodian that takes care of the sacred territories and the development of livelihoods ingrained, dependent and respectful of nature. In 2015, the African Commission on Human and People's Rights, acknowledging the critical role of sacred sites and with the input of nature custodians, adopted resolution 372 [33] for the protection of sacred natural sites in the African continent [34]. In Asia as well there is a deep-rooted tradition amongst indigenous populations existing in almost every corner. The Asian Indigenous People's Pact as a representative of indigenous populations from 14 countries is committed to achieving the "integrity of the environment" and to "enhance the sustainable resource management systems of indigenous peoples." [35].

There are many examples of indigenous populations that regard nature with deep respect and have a strong sense of belonging to it. Much of the indigenous knowledge and livelihoods that have achieved harmony with nature match up with modern notions of nature conservation [36], and their effectiveness in being in balance with nature can be explained by modern science. Their knowledge and experience on how to promote the rights of nature is being heard in international forums and should be heard as part of the legislative process to include the rights of nature in our legislations.

3. Can a River Stand in Court?

Worldwide legal systems are gradually introducing the possibility of granting rights to nature to stand in court for protection. Examples of court rulings applying said legislation and recognizing rights to nature have started to emerge a little more than five years ago. These rulings have granted rights to rivers across the world in different terms. There is no explanation about why rulings have been more prevalent regarding the rights of rivers than those of other ecosystems. The first ruling recognizing the rights of nature was regarding a river, and the existence of this previous jurisprudence could have provided foundations for other judges to rule in the same way. The nature of rivers as a distinct mass of water elapsing across terrain with a quasi-permanent shape and presence may make it easier to legally define a river as an "object" that can become a "subject" with rights. To date, rivers have been recognized as holding rights by a court ruling in Ecuador, India, New Zealand, and Colombia. These cases are the first judicial attempts to apply legislation that recognizes the rights of nature or to set precedence in recognizing such rights.

The first ruling was delivered in Vilcabamba, Ecuador. A public contractor started building a road next to the Vilcabamba River using dynamite and heavy machinery and depositing rocks and other construction materials in the river banks. The accumulation of these materials caused floods along the river and polluted the waters. After some affected citizens brought this case to the courts, the river's right to stand in court was admitted and those citizens representing the river continued in the process. The judge determined that the rights of nature had been violated—more specifically nature's right "to exist, to be maintained and to the regeneration of its vital cycles, structures and functions." This legal sentence did not stop the construction of the road, however. Instead the court ruled that the contractor should follow a set of environmental guidelines and recommendations that the Ministry of Environment had issued following a previous legal complaint against the road construction. It was then, on 30 March 2011 [37], that the rights of nature were recognized by a court for the first time. The ruling recognized the plaintiff's right to sue on the basis of Article 71 of the constitution, which establishes every citizen or nation's right to demand the authorities the compliance with the rights of nature. The ruling recognizes the rights of nature as a constitutional right to be observed and emphasizes that every citizen can defend such rights in court when violated. However, it does not further elaborate on when nature should hold locus standi per se. The court applies the precautionary principle deeming necessary to order the halt of any construction "until it is objectively demonstrated that there is no likelihood or danger" of environmental damage. Finally, to defend the construction works, the provincial governments alleged that respecting the rights of nature would mean the violation of the local's human right to development. To this allegation, the court responded that both rights are recognized by the constitution and should be pondered in the light of the constitutional principles. For this specific case, the court concluded that these rights are not colliding since the road can still be constructed while respecting nature's rights [38].

In New Zealand, members of the indigenous Maori tribes have disputed with the Crown the status of the Whanganui River for the last 140 years in the framework of the interpretation of the Treaty of Waitangi, a treaty declaring British Sovereignty in 1840 and defining Maori's land ownership, generally considered the founding document of New Zealand as a nation. Despite this, many Māori feel that the Crown did not fulfill its obligations under the Treaty and have presented evidence of this before sittings of the Waitangi Tribunal. In 2014, a settlement was finally reached [39] that would grant the river its own legal identity, with the rights, duties, and liabilities of a legal person. By this settlement, "the river becomes an entity in its own right, Te Awa Tupua" [39]. This settlement was turned into the Te Awa Tupua Act in 2017 by which the Whanganui becomes a legal person that will be able to be represented in court proceedings [40] and would have two guardians, one from the Crown and one from the Whanganui iwi [41] (see Part 2, Article 14, of the Te Awa Tupua 2017 Act: Te Awa Tupua is a legal person and has all the rights, powers, duties, and liabilities of a legal person). This has been so far the clearest legal reference to the way the rights of nature should be delineated and should be exercised. In this case, the Act, beyond declaring that nature has rights, explicitly grants legal personality to an entity within nature, i.e., the Whanganui River. It goes even further by naming those who should legally represent the river in court. The act makes a reference to the "Whanganui Iwi standing" (in Part 3, Subpart 2, named Ko au te Awa, ko te Awa ko au—Whanganui Iwi standing). It specifies that, for the purposes of the Resources Management 1991 Act, the trustees "are entitled to lodge submissions on a matter [...] affecting the Whanganui River" and are "recognized as having an interest [...] greater than any interest in common with the public generally." Given the novelty of this declaration, it will be a matter of time to see how these norms applied to practical matters.

Almost at the same time as the Te Awa Tupua Act was made official, the Uttrakhand High Court in India recognized that both the Ganges and its main tributary, the Yamuna, as well as "all their tributaries, streams, every natural water flowing with flow continuously or intermittently of these rivers" would be "legal and living entities having the status of a legal person with all corresponding rights, duties and liabilities" [42,43]. The case was brought to court when officials complained that the governments of Uttarakhand and Uttar Pradesh states were not cooperating with the federal government to set up a panel to protect river Ganges. The ruling [43] mentioned New Zealand's decision to recognize the Whanganui River as an ancestor and appointed legal custodians that would be the ones in charge of protecting the rivers (Paragraph 19 states: "The Director NAMAMI Gange, the Chief Secretary of the State of Uttarakhand and the Advocate General of the State of Uttarakhand are hereby declared persons in loco parentis as the human face to protect, conserve and preserve Rivers Ganga and Yamuna and their tributaries"). The court draws on the Supreme Court's jurisprudence regarding personhood for Hindu deities and reaffirms that Hindu deities as juridical persons are to be managed by those entrusted with the possession of their property. The court bases its decision on the need to protect the recognition and the faith of society given that both of these rivers "support and assist both the life and natural resources [...] of the community." The court regrettably did not elaborate on what the implications of such a declaration of rights would be, since the main focus of this ruling was actually on the nature of Indian federalism and the water management duties of federal and state governments and not on the rights of nature per se [44]. As aforementioned, three representatives are declared persons in loco parentis, as the persons in charge to protect the river. However, it is not discussed whether the river will hold locus standi whenever damaged or only under specific circumstances could these guardians defend the river's rights in court. This same court ruled in April of the same year that Himalayan glaciers Gangotri and Yamunotri are legal persons. However, the Indian Supreme Court later overturned both rulings [45] after the state of Uttrakhand argued that the ruling could lead to complicated legal situations given that the consequences of providing rights to these rivers were not clearly defined. The case has not been settled, as the petitioner intends to appeal [46], but it is reflective of many of the questions that are raised by those opposing the legal doctrine of the rights of nature. Uncertainty is certainly a challenge to overcome.

Also based on the need to protect the river from human activity was the Atrato River ruling in Colombia. Illegal mining activities near the Atrato River and its tributaries were polluting the river and damaging the livelihoods and health of those living in the area. Given the situation, the Center of Studies for Social Justice "Tierra Digna" [47] demanded the Government action to stop these activities and to protect the river. After this request was denied by the government, the case was brought to court. The judges noted the existence of a "serious violation of the fundamental rights to life, health, water, food security, the healthy environment, the culture and the territory of the ethnic communities that inhabit the Atrato River basin and its tributaries, attributable to the Colombian State entities." As a result, the court ordered that the river Atrato, its tributaries, and its basin have the right to be protected, preserved, and restored by the State and the communities. To safeguard these interests, the court mandates the government to appoint two representatives of the river, one would be a member of the community and the other a member of the government. Similarly, in this case, the river is provided with legal personhood and with representatives. However, when the river would have locus standi to be defended against any harm is unclear and has been left to be decided on a case-by-case basis.

The existence of this legal precedence is sparking actions for the recognition of the rights of nature in other parts of the planet. In fact, recently in September of 2017, Jason Flores-Williams (a lawyer in Denver) filed a suit in the Colorado Federal District Court seeking to hold the state of Colorado liable for the violation of the river's right to exist, flourish, and regenerate. The plaintiff in the suit is the river ecosystem, and, because the river itself cannot appear in court, Deep Green Resistance filed the suit as a "next friend" of the river (an individual who acts on behalf of another individual who does not have the legal capacity to act on his or her own behalf) [48]. The court has not ruled anything thus far but it is expected that similar cases will start sprouting all over the world. The movement supporting the rights of nature, and especially of rivers, to stand in court has found ground and encouragement on the existence of these precedents. Both the existence of political declarations and actual legislation make more plausible the idea of supporting the introduction of the rights of nature in other legal systems, and the existence of courts' rulings are setting a precedent by materializing the abstract idea of rights of nature into enforceable verdicts.

4. Conclusions and Challenges Moving Forward

Introducing an earth-centered paradigm in our legal systems and providing nature with the right to stand in court is only the first step towards the actual integration and enforcement of the rights of nature. In this process, there are still roadblocks, dilemmas, and unresolved questions.

First, it is not yet clear what the legal implications and meaning of providing legal personhood to nature might be. The definition of legal personhood [49] varies across legal systems and it has even been an issue of dispute among academics as the rights, duties, and powers to be exercised by a legal person are not easily definable by the law. Critics of the doctrine of the rights of nature have expressed concern over the attribution of legal personhood to nature as a source of legal uncertainty. This uncertainty is even more accentuated when it comes to defining when nature holds locus standi and on what basis. The debate remains whether the traditional theories that define both legal personhood and locus standi need to become more flexible to adapt to new paradigms and the way society perceives animals and nature. There is still a great deal of confusion in this sense, even within the legal spheres. For example, the India Supreme Court suspended the court order that declared the Ganges and Yamuna rivers to be "legal persons," arguing that rivers cannot be considered as "living entities" [50]. However, being a "living entity" should not be the focus of the legal debate. Especially since our current legal theory and practice support a system where businesses and governments go to court on a daily basis even though they are not "living entities." While enough evidence could be provided both in favor and against the fact that ecosystems are living beings, this discussion should be held outside of the courts. Legally speaking, a unified argument that ecosystems can be provided legal personhood similar to the one provided to private entities and governments would overall contribute to a better and unified understanding of the objective of the rights of nature doctrine, providing nature with the right to stand in court. This process to redefine legal personhood should also include an attempt to define whether, and under which circumstances, nature holds locus standi. Revisiting the debates that human rights experts have held could offer a good start in the process, even though different legal system requirements will have to be taken into account. Finally, even if the meaning of both legal personhood and locus standi for nature are clarified, further cogitation and plausible arguments on the value added to providing rights to nature will be needed to foster the expansion of the rights of nature. For some, it is clear that the traditional environmental legislation approach, in which nature is a passive being protected by the law, is not working. From a deep ecology perspective, our current anthropocentric approach should be shifted to have the Earth at the center of the system if we want to achieve an actual sustainable future. However, the question still remains whether providing rights to nature is the way to go.

Second, once a ruling has recognized certain rights to nature, it must be complied with. However, the enforcement of the rulings has proved challenging given the lack of precedent and of compliance mechanisms to protect nature once its rights have been recognized. For instance, in Vilcabamba (Ecuador), even though the ruling recognized the rights of the river and condemned the local government to apply a set of mechanisms to protect the integrity of the river, the plaintiffs noted how the enforcement of the ruling proved very challenging after the legal victory. As a matter of fact, while the ruling arranged that the Provincial Government of Loja (Gobierno Provincial de Loja) would deliver a remediation plan within thirty days, many months passed before the plan was submitted to the Ministry of Environment for approval. Moreover, while the local government partially complied with the ruling, the plaintiffs complained that they had to pay for the execution of certain works that would protect their property and the river because the local government would not deliver on its duties [51]. In the case of the Atrato River, the ruling ordered the creation of a "Commission of Guardians of the Atrato River" within three months of the ruling (the commission should count on two designated guardians and an advising team with members from the Humboldt Institute and WWF Colombia, both organizations with prior experience in the protection of other rivers in Colombia). While the ruling is from November 2016, the panel was only recently formed in October of 2017 [52]. Other uncertainties regarding the meaning and enforcement of the ruling were brought up when the state of Uttrakhand brought the ruling to India's Supreme Court seeking, amongst other things, clarification on whether the newly appointed custodians of the rivers or the state government were liable to pay damages (declarations of the Highest court in the Himalayan state of Uttarakhand). Trying to prevent this problem, recent rulings such as the one on the Atrato River (Colombia) have attempted to provide a more detailed ruling to ease its enforcement. The ruling defined in detail the institutions that would be in charge of watching the ruling's compliance and diversify this task rather than giving it to one institution. In this case, the Attorney General would be in charge of it, with the support of the Communities, the Ombudsman, and the General Comptroller's Office [53]. Future court rulings and national legislation should follow these steps and define more carefully how the recognition of the rights of nature have to be enforced once recognized.

Third, if this legal approach was actually adopted, the question of how much should we give up in terms of development in order to respect the rights of nature remains. Several different philosophical approaches provide different answers to this question. According to the Sumak Kawsay or "living well" approach to development, we only need to develop enough to live well and in harmony with nature. This perspective challenges the current approach to development, associated with a perpetually increasing pattern of natural resource exploitation and growth. In an earth-centered paradigm, the rights of humans do not clash with the rights of nature because they have the same objective: to live in harmony. Therefore, if this paradigm was followed, the human approach to development would also shift. However, in practical terms, establishing the right balance between human development and the respect of nature's rights will prove challenging for the courts. Many fear that a shift in paradigm could lead to mountains or forests to sue over the depletion of their natural resources or the pollution of ecosystems and that this could stop the development of housing complexes, or roads, or other infrastructures. This process, however, already happens on a daily basis, when courts make decisions to allow or restrain corporations and governments' actions that affect a community. Nonetheless, we are still far from an ultimate answer in this regard.

So far, the governments of Bolivia [54] and Ecuador [55], which have strongly pushed for the declaration of the rights of nature at an international level, have adopted a very pragmatic approach within the country [56]. While their legal systems have adopted the rights of nature doctrine, these countries' extractive industries are blooming to the detriment of the ecosystems surrounding them. Social conflicts with the indigenous population attempting to protect nature in these areas are still prevalent, and there are ongoing debates on the tradeoffs of natural protection over resource extraction and the benefits of one or another at a national level. Nevertheless, the fact that, in these countries, the rights of nature are contained in their constitutions at least presents an opportunity to rethink and re-politicize the environmental debate [57].

All in all, the recognition of the rights of nature is still a very incipient movement within formal legal systems. It is a movement that brings along many uncertainties, but also the potential to fully develop and become the rule instead of the exception. When rethinking our current legal system and attempting to introduce an Earth-centric paradigm and its enforcement in court, these challenges should be kept in mind. History has proven that law often lags behind social change. As Leimbacher said, "legal standing for nature is nothing but a consequential continuation of a century-long process of expansion of the group of legal subjects." [6]. This is, arguably, the process that we are witnessing right now. The legal doctrine of the rights of nature is still being developed and changes in our paradigm are still underway. However, the rights of nature are here to stay. As the planet strives to achieve a more sustainable way of living, the rights of nature will offer a legal tool to regulate our relationship with nature from a different and more harmonious perspective. The court rulings, regulations, and political declarations discussed in this paper, even though still filled with uncertainties, play an important role in confirming and materializing the new values of deep ecology that are slowly growing within society. Legal uncertainties must be addressed by jurists, but reaching the right balance and building a robust system will only be reached through a trial and error process. What has been described in this review are the foundations over which we are building a new paradigm and the first steps towards a robust legal approach to recognizing the rights of nature, both in theory and in legal practice.

Acknowledgments: The author wrote this article under the sponsorship of the Fulbright program and Fundacion Ramon Areces. However, these sources did not cover the costs to publish in open access.

Conflicts of Interest: The authors declare no conflict of interest.

References

1. Stone, C.D. Should Trees Have Standing?—Towards Legal Rights for Natural Objects. *South. Calif. Law Rev.* **1972**, *45*, 450–501.
2. Neimark, P.; Mott, P.R. *The Environmental Debate: A Documentary History*, 2nd ed.; Grey House Publishing: Aminia, NY, USA, 2011; ISBN 1592376762.
3. Dred Scott, v. Sandford. 60 U.S. (19 How.) 396, 404–405 (1856). In Bailey v. Poindexter's Executor, 56 Va. (14 Gratt.) 132, 142–143 (1858), Superseded by Constitutional Amendment, U.S. Const. amend. XIV. Available online: https://www.loc.gov/rr/program/bib/ourdocs/DredScott.html (accessed on 1 October 2017).
4. Wise, S. Animal rights, one step at a time. In *Animal Rights: Current Debates and New Directions*; Oxford Scholarship Press: Oxford, UK, 2005; p. 26.
5. Peter, S. *Practical Ethics*; Cambridge University Press: Cambridge, UK, 1979; Chapter 3; pp. 55–82, 151, ISBN 0521229200.
6. Leimbacher, H. Gender and Nature in Comparative Legal Cultures. In *Comparing Legal Cultures*; Nelken, D., Ed.; Routledge: New York, NY, USA, 2016; Chapter 8; p. 146. ISBN 9781855218987.
7. Nedelsky, J. Reconceiving Rights as Relationship (1993), Volume 1, Issue 1 (1993). Available online: https://ssrn.com/abstract=2045687 (accessed on 11 February 2018).
8. Universidad Española a Distancia. Teoria del Derecho. Available online: http://ocw.innova.uned.es/ocwuniversia/teoria-del-derecho/teoria-del-derecho/resumenes-1/tema-10-persona-y-personalidad-juridica-capacidad-juridica-y-capacidad-de-obrar (accessed on 7 January 2017).
9. Korsgaard, C.M. *Personhood, Animals, and the Law*; Cambridge University Press, 2013; Volume 12, pp. 25–32. Available online: https://doi.org/10.1017/S1477175613000018 (accessed on 1 October 2017).
10. Kurki, V.A. Revisiting Legal Personhood. June/July 2016. Available online: http://www.uef.fi/documents/300201/0/Kurki++Revisiting+legal+personhood.pdf/56e99525-ba38-4c05-8034-3505d52d84a0 (accessed on 1 October 2017).
11. Dyschkant, A. Legal Personhood: How We Are Getting It Wrong. Illianois Law Review. Volume 2015, p. 2075. Available online: https://illinoislawreview.org/wp-content/ilr-content/articles/2015/5/Dyschkant.pdf (accessed on 30 December 2017).
12. Brei, A.T. Rights & Nature. *J. Agric. Environ. Ethics* **2013**, *26*, 393–408.
13. Tanasescu, M. Local, National, and International Rights of Nature. In *Environment, Political Representation, and the Challenge of Rights*; Palgrave Macmillan: London, UK, 2016; pp. 107–128. ISBN 978-1-349-55977-0.
14. Legal Text of the Ecuadorian Constitution of 2008. Available online: http://www.asambleanacional.gob.ec/sites/default/files/documents/old/constitucion_de_bolsillo.pdf (accessed on 22 October 2017).
15. Legal Text of the Bolivian Constitution of 2009. Available online: http://www.harmonywithnatureun.org/content/documents/159Bolivia%20Consitucion.pdf (accessed on 20 October 2017).
16. Ecuadorian Constitutional Court Ruling Number 017-12-SIN-CC Case Number 0033-10-IN. Available online: http://doc.corteconstitucional.gob.ec:8080/alfresco/d/d/workspace/SpacesStore/cccd3c44-11af-48bd-85f7-148c01ccfd36/0033-10-IN-sent.pdf?guest=true (accessed on 05 October 2017).
17. Bolivian Law 071, Ley de Derechos de la Madre Tierra, 21 December 2010. Available online: http://www.harmonywithnatureun.org/content/documents/158Bolivia%20Ley%20071.pdf (accessed on 20 October 2017).
18. Barie, C.G. Nuevas Narrativas Constitucionales en Bolivia y Ecuador: El Buen Vivir y los Derechos de la Naturaleza. *Latinoamérica. Revista de Estudios Latinoamericanos* **2014**, *59*, 9–40. [CrossRef]
19. El Potosi. Denuncian a Evo ante el Tribunal de Derechos de la Naturaleza. 9 November 2017. Available online: http://elpotosi.net/nacional/20171109_denuncian-a-evo-ante-el-tribunal-de-derechos-de-la-naturaleza.html (accessed on 15 November 2017).

20. General Assembly Resolution 63/278, International Mother Earth Day, A/RES/65/164, 15 March 2011. Available online: un.org/en/ga/president/65/initiatives/Harmony%20with%20Nature/A-RES-65-164.pdf (accessed on 1 October 2017).

21. General Assembly Resolution 07/208, Harmony with Nature, A/RES/70/208, 14 December 2012. Available online: un.org/ga/search/view_doc.asp?symbol=A/RES/70/208 (accessed on 1 October 2017).

22. General Assembly Resolution, Sustainable Development and Sustainable Development Goals, A/68/948, 7 July 2014. Available online: http://www.g77.org/doc/A-68-948(E).pdf (accessed on 1 October 2017).

23. IUCN, Programme 2017–2020. September 2016. Target 6. Available online: https://static1.squarespace.com/static/55914fd1e4b01fb0b851a814/t/57f6894f440243a1628b3690/1475774800092/IUCN+Programme+2017-2020-FINAL+APPROVED.pdf (accessed on 20 November 2017).

24. Argentinian Senate. First Debate on the National Senate on the Rights of Nature. 7 July 2015. Available online: http://www.senado.gov.ar/prensa/13264/noticias (accessed on 30 October 2017).

25. Municipal Chamber of Sao Paolo. Projeto de Emenda a Lei Organica 04-00005/2015 do Vereador Eduardo Tuma (PSDB). Available online: http://cmspbdoc.inf.br/iah/fulltext/projeto/PLO0005-2015.pdf (accessed on 30 October 2017).

26. Mercado, J. Legal Recognition of the Sacredness of the Earth: Rights of Nature. Pachamama Alliance. 28 October 2015. Available online: https://www.pachamama.org/blog/legal-recognition-of-the-sacredness-of-the-earth-rights-of-nature (accessed 30 October 2017).

27. Mumta Ito. ECI Project—A European Citizens Initiative for the Rights of Nature (Working Draft). Available online: https://docs.google.com/document/d/1BywdKZXULM4eIkMv5BuIpksRjAjwMPKTCHJ10tRPoZw/pub (accessed on 1 October 2017).

28. Kiana, H. "Los Derechos de la Naturaleza: Las Filosofias Indigenas Estan Reformulando la Ley". Intercontinental Cry, January 2017. Available online: https://intercontinentalcry.org/es/los-derechos-de-la-naturaleza-las-filosofias-indigenas-estan-reformulando-la-ley/ (accessed on 15 November 2017).

29. Zaffaroni, E.R. La Pacha Mama y el Humano. 2012. Available online: http://www.cuspide.com/Libro/9789505639250/La+Pachamama+Y+El+Humano (accessed on 11 February 2018).

30. Pachamama Alliance. Sumak Kawsay: Ancient Teachings of Indigenous Peoples. Available online: https://www.pachamama.org/sumak-kawsay (accessed on 20 November 2017).

31. Herold, K. The Rights of Nature: Indigenous Philosophies Reframing Law. Deep Green Resistance News Servic. Available online: https://intercontinentalcry.org/rights-nature-indigenous-philosophies-reframing-law/ (accessed on 20 October 2017).

32. Rubiano, M.P. "Si los Bosques Siguen en Pie es por Nosotros": Indígenas en la Cumbre del Cambio Climático. *El Espectador.* 15 November 2017. Available online: https://www.elespectador.com/noticias/medio-ambiente/si-los-bosques-siguen-en-pie-es-por-nosotros-indigenas-articulo-723281 (accessed on 20 November 2017).

33. Afican Commission on Human and People's Rights. ACHPR/Res 372 (LX) 2017. Resolution on the Protection of Sacred Natural Sites and Territories. Available online: http://www.achpr.org/sessions/60th/resolutions/372/ (accessed on 10 November 2017).

34. Wilton, F. Respect Sacred Natural Sites to Guarantee Human Rights, Says New African Commission Resolution. IUCN, 11 October 2017. Available online: https://www.iucn.org/news/commission-environmental-economic-and-social-policy/201710/respect-sacred-natural-sites-guarantee-human-rights-says-new-african-commission-resolution (accessed on 20 October 2017).

35. Verschuren, B.; Wild, R.; McNeely, J.; Oviedo, G. *Sacred Natural Sites, Conserving Nature and Culture*; Earthscan: London, UK; Washington, DC, USA, 2010; p. 22. ISBN 978-1-84971-166-1.

36. Asia Inigenous Pact Mission and Vision. Asia Indigenous Pact Official Website. Available online: https://aippnet.org/about-us/ (accessed on 30 October 2017).

37. UNEP. Indigenous People and Nature: A Tradition of Conservation. 26 April 2017. Available online: http://web.unep.org/stories/story/indigenous-people-and-nature-tradition-conservation (accessed on 20 October 2017).

38. Ruling by the Ecuadorian Sala Penal de la Corte Provincial. Protection Action. Ruling Number No. 11121-2011-0010. Casillero N0. 826. 30 March 2011. Available online: http://consultas.funcionjudicial.gob.ec/informacionjudicial/public/informacion.jsf (accessed on 10 October 2017).

39. El Correo. Jurisprudencia Ecuatoriana sobre Derechos de la Naturaleza. 8 June 2011. Available online: http://www.elcorreo.eu.org/IMG/article_PDF/Jurisprudencia-Ecuatoriana-sobre-Derechos-de-la-Naturaleza_a20229.pdf (accessed on 10 October 2017).

40. Williams, J. Te Awa Tupua. Kokiri: Raumati, 2016; pp. 28–31. Available online: https://www.tpk.govt.nz/en/mo-te-puni-kokiri/kokiri-magazine/kokiri-33-2016/te-awa-tupua (accessed on 20 October 2017).

41. Davison, I. Whanganui River Given Status of a Person under Unique Treaty of Waitang Settlement. *New Zealand Herald*, 15 March 2017. Available online: http://www.nzherald.co.nz/nz/news/article.cfm?c_id=1&objectid=11818858 (accessed on 10 October 2017).

42. Roy, E.A. New Zealand River Granted Same Legal Rights as Human Being. *The Guardian*, 16 March 2017. https://www.theguardian.com/world/2017/mar/16/new-zealand-river-granted-same-legal-rights-as-human-being (accessed on 10 October 2017).

43. Safi, M. Ganges and Yamuna Rivers Granted Same Legal Rights as Human Beings. *The Guardian*, 21 March 2017. Available online: https://www.theguardian.com/world/2017/mar/21/ganges-and-yamuna-rivers-granted-same-legal-rights-as-human-beings (accessed on 30 December 2017).

44. Mohd, S.V. State of Uttarkhand and Others (Writ Petition (PIL) No. 126 of 2014. 20 March 2017. Available online: http://lobis.nic.in/ddir/uhc/RS/orders/22-03-2017/RS20032017WPPIL1262014.pdf (accessed on 30 December 2017). Paragraph 19.

45. Vrinda, N. Indian Court Recognizes Rivers as Legal Entities. ICONnect, International Journal of Constitutional Law Blog. Available online: http://www.iconnectblog.com/2017/06/indian-court-recognizes-rivers-as-legal-entities/ (accessed on 10 October 2017).

46. BBC. India's Ganges and Yamuna Rivers Are 'Not Living Entities'. 7 July 2017. Available online: http://www.bbc.com/news/world-asia-india-40537701 (accessed on 30 October 2017).

47. RTE. Indian Supreme Court Rules Rivers Are Not People. 7 July 2017. Available online: https://www.rte.ie/news/2017/0707/888557-india-rivers/ (accessed on 30 October 2017).

48. Colombia Constitutional Court Ruling T-622 of 2016, Expediente T-5.016.242. Available online: https://justiciaambientalcolombia.org/2017/05/07/sentencia-rio-atrato/ (accessed on 20 November 2017).

49. Turkewitz, J. Corporations Have Rights, Why Shouldn't Rivers. *The New York Times*, 26 September 2017. Available online: https://www.nytimes.com/2017/09/26/us/does-the-colorado-river-have-rights-a-lawsuit-seeks-to-declare-it-a-person.html?_r=0 (accessed on 20 November 2017).

50. Quintana Adriano, E. Natural Persons, Juridical Persons and Legal Personhood. *Mex. Law Rev.* **2015**, *8*, 101–118. [CrossRef]

51. The Times of India. Supreme Court Stays Uttarakhand High Court's Order Declaring Ganga and Yamuna 'Living Entities'. 7 July 2017. Available online: https://timesofindia.indiatimes.com/india/supreme-court-stays-uttarakhand-high-courts-order-declaring-ganga-and-yamuna-living-entities/articleshow/59489783.cms (accessed on 30 November 2017).

52. Suarez, S. Defendiendo la Naturaleza: Retos y Obstáculos en la Implementación de los Derechos de la Naturaleza Caso río Vilcabamba. FES Energia y Clima 2013. Available online: http://library.fes.de/pdf-files/bueros/quito/10230.pdf (accessed on 10 October 2017).

53. Colombia Attorney General Bulletin 829. 15 October 2017. Available online: https://www.procuraduria.gov.co/portal/Procuraduria-integra-panel-de-expertos-para-proteger-rio-Atrato.news (accessed on 20 October 2017).

54. Tierra, D. Todas y Todos Somos Guardianes del Atrato. Available online: http://tierradigna.org/pdfs/SomosGuardianesDelAtrato.pdf (accessed on 30 October 2017).

55. Schilling-Vacaflor, A. *Contestations over Indigenous Participation in Bolivia's Extractive Industry: Ideology, Practices, and Legal Norms*; GIGA (Working Paper), No. 254, September 2014; GIGA German Institute of Global and Area Studies: Hamburg, Germany, 2014.

56. Steven, C.; Winterbottom, R.; Reytar, K.; Strong, A. Ecuador Shows Why Communities and the Climate Need Strong Forest Rights. World Resources Institute, 19 September 2014. Available online: http://www.wri.org/blog/2014/09/ecuador-shows-why-communities-and-climate-need-strong-forest-rights (accessed on 20 November 2017).

57. Lalander, R. Rights of Nature and the Indigenous Peoples in Bolivia and Ecuador: A Straitjacket for Progressive Development Politics? *Iberoam. J. Dev. Stud.* **2014**, *3*, 148–172.

resources

MDPI

Article

Recognition of Barkandji Water Rights in Australian Settler-Colonial Water Regimes

Lana D. Hartwig [1,2,*], Sue Jackson [1] and Natalie Osborne [2]

[1] Australian Rivers Institute, Griffith University, Nathan Campus, 170 Kessels Road, Nathan, QLD 4111, Australia; sue.jackson@griffith.edu.au

[2] School of Environment and Science, Griffith University, Gold Coast Campus, Parklands Drive, Southport, QLD 4222, Australia; n.osborne@griffith.edu.au

* Correspondence: lana.hartwig@griffithuni.edu.au; Tel.: +61-405-450-453

Received: 25 November 2017; Accepted: 14 February 2018; Published: 24 February 2018

Abstract: The passage of the *Native Title Act 1993* (Cth) brought with it much anticipation—though in reality, quite limited means—for recognizing and protecting Aboriginal peoples' rights to land and water across Australia. A further decade passed before national and State water policy acknowledged Aboriginal water rights and interests. In 2015, the native title rights of the Barkandji Aboriginal People in the Australian State of New South Wales (NSW) were recognized after an eighteen-year legal case. This legal recognition represents a significant outcome for the Barkandji People because water and, more specifically, the Darling River, or *Barka*, is central to their existence. However, the Barkandji confront ongoing struggles to have their common law rights recognized and accommodated within Australian water governance regimes. Informed by literature relating to the politics of recognition, we examine the outcomes of government attempts at Indigenous recognition through four Australian water regimes: national water policy; native title law; NSW water legislation; and NSW water allocation planning. Drawing from the Barkandji's experiences in engaging with water regimes, we analyze and characterize the outcomes of these recognition attempts broadly as 'misrecognition' and 'non-recognition', and describe the associated implications for Aboriginal peoples. These manifestations of colonial power relations, whether intended or not, undermine the legitimacy of state water regimes because they fail to generate recognition of, and respect for, Aboriginal water rights and to redress historical legacies of exclusion and discrimination in access to water.

Keywords: Indigenous peoples; Aboriginal peoples; native title; politics of recognition; Indigenous water rights; water governance; water planning; New South Wales; Darling River

1. Introduction

> ' ... *when the government came out and gave us our native title rights, it was recognition we are Barkandji People ... They gave us our native title rights and took our water and that's the most valuable thing: the water and the land.*' (Barkandji Prescribed Body Corporate Director D, 13 February 2017)

Much of Australia's colonial wealth has been built on exploiting water resources for irrigation, mining and urban water supply. This exploitation has involved encroachment of Aboriginal and Torres Strait Islander peoples' rights and interests in water, in addition to rights to land [1,2]. The existence of Aboriginal customary water law and management systems prior to British occupation in 1788 was denied by European settlers for more than two centuries, as was Aboriginal title to land [3]. In 1992, the *Mabo* decision of Australia's High Court precipitated a 'judicial revolution' [4] that radically revised the explanation of the law that had evolved from foundational colonial acts of occupation. As a consequence of *Mabo*, Australian law responded with the passage of Commonwealth legislation in

1993, the Native Title Act (NTA hereafter), which recognizes: that there were legal systems in place at the time of European occupation; that Indigenous peoples' rights to land survived colonization; and that a form of native title could exist in situations in which it had not been extinguished [5].

With the passage of the NTA, a means was provided for Aboriginal and Torres Strait Islander peoples to legally claim unalienated land in those places where they could prove continuity of customs and traditions and uninterrupted connection to customary estates. The scope of the NTA was defined to include rights over waters located within traditional estate boundaries. It confirmed Crown ownership of water and minerals, while guaranteeing rights to customary use of resources for sustenance (hunting, gathering and fishing). In addition, a right to protect sites or areas of cultural significance that include waters has been recognized as a native title right. A native title right to take and use water for commercial purposes is yet to be recognized by the courts.

It took Australian water policymakers a further decade to acknowledge Indigenous water interests in the series of water governance reforms called National Water Initiative (NWI) of 2004. These reforms included the creation of legally tradeable and transferable water use entitlements, pricing of water, and, in more recent legislative change, the re-allocation of substantial amounts of water from agriculture to the environment in Australia's south-east. These neoliberal reform initiatives included recommendations to improve Aboriginal access to water for 'traditional, cultural, spiritual and customary' purposes and for increased Aboriginal participation in water planning and management [6]. In creating these legislative and policy institutions, Australian governments deployed the concept of customary practice to signal culturally distinctive forms of water rights (cf. [7]). Policy and law reform has since admitted only 'very limited, narrowly prescribed, and externally defined spaces for Indigenous Peoples to influence decisions about water use and management' [8].

As a result of the historical legacy of Australian colonization, the current distribution of water entitlements to Aboriginal peoples remains transparently unjust [3]. This injustice stems from the fact that Aboriginal peoples had free access to use and benefit from water until their lands and waters were taken without consent or compensation. As of 2013, there have been at least 34 land redistribution measures introduced by Australian governments since 1966 to redress Aboriginal land dispossession, which together have returned over a third of Australia's landmass with varying degrees of control and influence to Aboriginal peoples [9]. By comparison, however, as of 2012, Aboriginal peoples held less than 0.01 per cent of Australia's water diversions and, as we will show in this article, recent government efforts to improve Aboriginal water access have had negligible effect on increasing Aboriginal-held water allocations [3]. Moreover, Aboriginal peoples continue to hold a weak legal position in Australia's water governance frameworks, which constrains their influence in water planning and the allocation or sharing of water resources [10,11]. In regions where Aboriginal people face a high degree of contestation and competition for water, such as in the south-east of Australia where much of the country's irrigated agriculture occurs, policy and legal frameworks fail to address their rights and interests, which are seen as outside of, or irrelevant to, the formal economy. Governments may issue water licences or permits to take and use water to other parties, and Aboriginal communities are denied any opportunity to object or to negotiate commercial benefits in relation to this third party access [12]. Commercial uses and benefits from the limited instances where Aboriginal communities are specially accorded water rights are also prohibited [3]. In addition, the diversion of water and over-allocation of entitlements has resulted in severe environmental degradation that jeopardizes Aboriginal peoples' lifeways [13].

Australia's settler-colonial water regimes have now espoused recognition of Aboriginal water rights and values for over 15 years. In this article, we aim to explore (a) the recognition-based regimes used by the state to acknowledge and accommodate Aboriginal water rights claims in New South Wales (NSW); (b) the nature and extent of water rights and control that is actually conferred to Aboriginal peoples under these water regimes; and (c) the consequent outcomes and implications. To do this, we draw on the struggle over water rights in the Australian State of NSW by Barkandji Traditional

Owners (a Traditional Owner is defined as an Aboriginal person who is a member of a local descent group, having certain rights and responsibilities in relation to an area of land, water or sea).

As we will elaborate, the Barkandji's water rights struggle (and indeed, this article) is not a comparison of the treatment of the Barkandji People and their water rights with other water users, such as local farmers. Instead, their struggle centers on the NSW Government on the one hand recognizing that the Barkandji People have particular rights to be included in decision making about and accessing water with the provisions associated with native title, while on the other hand also failing to acknowledge, honor, or uphold those rights within its water regimes. We examine four regimes that determine the nature and character of Aboriginal water rights in NSW, including the Barkandji's: national water policy, native title legislation, NSW water legislation, and NSW water allocation planning. Drawing from post-colonial literature concerned with the politics of recognition, we problematize the forms of state recognition that underpin these regimes. Struggles for recognition reflect 'deeper, more fundamental material and structural inequalities that block equal participation' [14] (p. 14); thus, we find recognition to be a useful lens for understanding the power relations and asymmetries that shape and underpin water governance. For the purposes of this article, 'the state' in a generic sense, is composed of the ruling governments, institutions and agencies that operate, govern and oversee society within that state's territory [15]. Therefore, 'the state' is not inherently synonymous with the nation state, but can also refer to self-governing political entities or jurisdictions within Australia such as at the Federal, State or Local Government levels. Additionally, the use of 'State' (capitalized) is used when describing States of Australia as stipulated under the Constitution of Australia, such as the State of NSW.

Our analysis suggests that under these current water regimes, states commit to recognizing Aboriginal peoples and their water rights via a set of institutional practices and that these attempts at recognition can have problematic outcomes that fall well short of progressive transformation. These processes require that the relationships and engagements Aboriginal peoples had, have, or seek to have with water, are made legible to—and thus, governable by—the state. The translation processes of making something legible such that it can be known and governed by the state without transforming its apparatuses or epistemologies distorts by flattening complexity, decontextualizing practices and relationships, and seeking to integrate knowledges and practices into existing frameworks without transforming those frameworks [16–18]. These processes reflect what might be called 'incommensurability' [19]. The ensuing outcomes can have the effect of misrecognizing Aboriginal rights to water by circumscribing and distorting them, or can completely ignore them through what we call non-recognition. It is our conjecture that these outcomes, resulting from settler state attempts at recognition, can serve to defer and diffuse more radical Aboriginal challenges to state sovereignty in that power and control remain firmly in the hands of the state [18]. This article aligns in some ways with the work by Behrendt and Thompson [1], who, as independent advisors to a NSW Government agency, analyzed and examined 'ways to recognize Aboriginal interests in NSW rivers' (p. 41) at the time most of the NSW-based water regimes examined in this article were implemented. Nearly 15 years later, this article and its findings build on and substantiate many of their early concerns about how the then-new water regimes might restrict and misrecognize Aboriginal water rights, showing how this has played out in a specific case.

This article is structured as follows. After briefly canvassing the literature relating to the politics of recognition, we establish our case study—the recent water rights struggle of the Barkandji People—and detail the methods used to inform this article. Next, four water regimes that generally shape the recognition of Aboriginal water rights (particularly within the State of NSW) are delineated, followed by an account of the Barkandji's attempts to have their rights and interests recognized through these regimes. We analyze recognition attempts by governments through these regimes, and argue that the outcomes of such efforts can be conceptualized as 'misrecognition' and 'non-recognition'. We further show that misrecognition and non-recognition both serve to undermine the legitimacy of state water regimes. We conclude with recommendations to redress these deficient outcomes.

2. The Politics of Recognition

Water is a mobile substance that is subject to multiple and sometimes competing demands, as well as increasing state control and development [20]. There is therefore often a need for Indigenous peoples dispossessed of their territories by colonial powers to seek state recognition of their legitimate authority over water, their normative constructs, and quotidian water use practices. In response to the civil rights movements of the 1960s, many liberal states developed mechanisms for recognizing these kinds of group-based claims of oppressed minorities, and granting them special rights (e.g., affirmative action and limited sovereignty) [21,22]. Recognition has since evolved as the leading framework through which to redress historical legacies and injustices of exclusion, racism and other forms of discrimination, and to enhance the freedom of Indigenous minorities. This is particularly so in settler states [23,24], where the logic of colonization and invasion ultimately seeks to secure access and complete control over territories and resources (including water) for the benefit of the settling colonies, and thus justifies the coercion, dispossession and elimination of Indigenous peoples [25,26].

Recognition proponents anticipated that the expansion of legal and cultural norms to Indigenous peoples would 'achieve greater equality of recognition as legal persons within a political community understood as legitimate and pre-existing' [27] (p. 6). Recognition politics has been envisaged as a 'philosophical and institutional remedy to matters of "historical injustice"' [24] (p. 20). For Ivison [28], historical injustice refers not only to

> acts of injustice that occurred in the distant past, but also how consequences of these injustices persist. These are enduring injustices—ones that continue to shape the conceptual, legal, political and institutional frameworks within which states and their citizens act. (p. 119)

Institutional expressions of recognition manifest in the 'political projects of reconciliation, multiculturalism and development, including the granting of land rights, constitutional recognition, or social, political or material entitlements' [27] (p. 1). The recognition of Indigenous water rights represents one such institutional expression [29].

The concept of recognition has been the subject of ongoing debates over the past three to four decades, with some taking issue with its tendency to dramatically simplify, reify and essentialize group characteristics, and its encouragement of intolerance and separatism [21]. Others point out the power and resource asymmetries present in forms and practices of recognition at both individual and societal levels [30]. Political theorists like Ivison [28] identify the disempowering paradox that lies at the heart of recognition politics:

> to seek recognition is to seek to be valued by others, which invites a critical evaluation of the beliefs and practices of the person (or peoples) making the claim. The "recogniser" thus exercises power over the "recognisee" in having the capacity to grant recognition. (p. 121)

Where states hand down single-directional, pre-determined offers of recognition like this, they reassert their presumed power and fail to discuss or listen to those demanding recognition, an approach to recognition that Tully has termed a monological orientation [30].

Within the literature on water rights and justice, Boelens [31] and Boelens et al. [32] argue that institutional recognition of Indigenous peoples similarly poses enormous conceptual challenges and important social and political consequences for those working on water justice projects. These challenges stem from the complexities associated with identifying, recognizing and formalizing diverse and dynamic Indigenous water rights and water claims by the state. Recognizing and practicing legal pluralism by states that are fundamentally hierarchical presents significant challenges [31], as does the possibility that these rights could be re-defined and possibly over-simplified to fit within the state's own frameworks [33].

Writing in response to North American settler colonialism, Indigenous scholars Simpson [24] and Coulthard [34] question the authority of settler states to 'recognize' Indigenous peoples, and their pre-existing and ongoing rights to govern themselves. They argue that relationships between

Indigenous peoples and (North American) settler states cannot be significantly transformed through recognition, and theorize alternative strategies that do not reinforce state dominance [34], including the option of 'refusal' [24]. Povinelli [18] suggests state recognition of Indigenous identity in this way generally 'supports and strengthens the nation and capital, not [I]ndigenous peoples' (p. 56). This is because this recognition is premised on a 'fantasy' of ancient laws and traditions that serves as 'a form of otherness that . . . does not violate the core subjective of social values of settler society' [18] (p. 65). Her work serves to illuminate the deep interconnections of how power operates and is configured through specific recognition approaches by settler states, namely those which emphasize Indigenous 'tradition' as one among many types of cultural difference within the multicultural state [18].

In Australia, where there has yet to be any form of treaty between any arm of the state and any group of Aboriginal peoples, arguably the extent to which Aboriginal people must submit to, or abide by, the 'recognition' or legitimization by the state is unclear and contestable [35]. As Australian Aboriginal scholar Watson [36] notes, 'many of us affirm our sovereignty as people who have never entered into consensual relations with any state or British Crown to surrender our international status' (p. 13). Notably in the context of Australia, where sovereignty is asserted by Aboriginal (and/or Torres Strait Islander) peoples it is usually in the context of desire for negotiation, or an agreement or a treaty-making process, rather than an assertion, for instance, of unilateral decision-making to the exclusion of the state. A very recent example of this is the *Uluru Statement from the Heart*, a declaration and recommendation from the First Nations Constitutional Committee that in 2017 asserted Indigenous sovereignty—their pre-existing and ongoing rights to govern themselves [37].

Fraser [21] argues that not all types and instances of recognition politics are equally pernicious, and Hunt [23] comments that 'strategically, it does not seem that outright rejection of all forms of recognition are politically viable' (p. 29). There may be symbolic and material benefits attached to some forms of recognition, and, although these structures may uphold asymmetrical colonial power relationships and systems of governance, in some contexts they might provide the only feasible means for Indigenous peoples to pursue secure access to their territories, including to water and waterways. This is particularly so in Australia, where, as mentioned, no arm of the state has ever attempted to gain the consent of Aboriginal peoples (through treaties, for instance) [35]. Thus, the tension between contesting the dominance of the settler-colonial state through decolonial politics, and securing some form of access, rights, and material benefits through state-based recognition processes, is particularly significant and warrants scrutiny.

In the analysis to follow, we review how the politics of recognition is playing out with respect to Aboriginal water rights in Australia, as illuminated by the experiences of the Barkandji People, a cluster of related tribes from NSW. Next we establish the historical and regional context of this case study.

3. Case Study and Methods

The Barkandji People's name derives from the *Barka* (the Darling River), meaning literally people of or belonging to the river [38]. Europeans called the *Barka* the Darling River in 1829 [39] after the then Governor of NSW, Sir Ralph Darling. Barkandji country is comprised of this major river system, together with the surrounding largely arid lands in most of far-western NSW between the Murray River in the south and parts of southern Queensland in the north [40]. European occupation of Barkandji country began as early as the 1830s, as settlers used the Darling River as a transport corridor [40] and imposed British riparian law to enable expansion of the pastoral frontier [41,42]. Like many other Aboriginal and Torres Strait Island groups across Australia, the Barkandji faced severe disruption to their way of life, having to contend with policies of displacement and relocation, even eradication. Many Barkandji were coerced into church missions and government reserves and were exploited by the regional pastoral economy [39,43]. (In Australia's colonial history, a reserve served as 'a place for the exclusive occupation by native tribes' (OED), while reserves were declared to manage Aboriginal peoples. Their declaration did not entail the grant of any rights to land to Aboriginal people and their use was paternalistically determined by colonial authorities.) As a consequence, Aboriginal peoples from this area (and other closely settled regions of south-east Australia) now have low rates of land ownership and experience other forms of socio-economic disadvantage [39]. These factors have left a legacy of 'low educational and employment outcomes, poor health and housing' [44] (p. 325), as well as spiritual, cultural, social, community, and familial impacts including intergenerational grief and loss [44,45]. These legacy issues are compounded by low availability of, and access to, support services and employment options [44], a common characteristic of Australia's regional and remote areas.

Seeking to have their unceded ancestral and customary rights and interests to these landscapes recognized by Australia's common law, the Barkandji submitted a native title claim soon after the introduction of the NTA. In October 1997 they lodged a claim to an area of lands and waters exceeding 128,000 km^2 (Figure 1). Nearly 18 years later, in June 2015, the Federal Court determined that the Barkandji held native title rights and interests to parts of the lands and waters within the area claimed [46]. While it is the largest claim of its kind in NSW [47], native title rights and interests were determined to be extinguished in the majority of the area [46]. More detail on the rights and interests recognized through their native title determination are provided in Section 4.2 below.

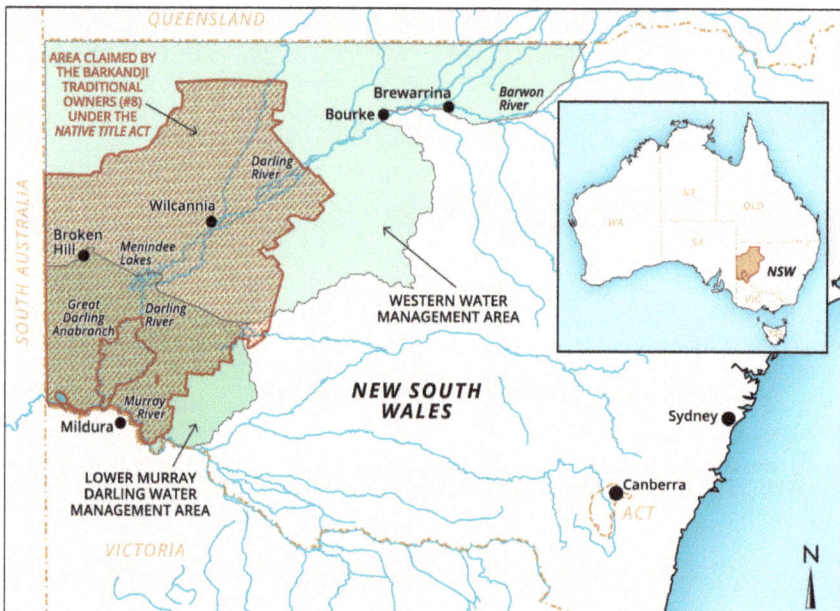

Figure 1. Map of western NSW showing the Barkandji Tradition Owners (#8) native title claim area and the water management areas it traverses. Note: Thick brown line denotes boundary of the area claimed by the Barkandji People. The lower reaches of the Darling River were excluded from this claim but surrounding lands were included.

Within the Barkandji native title claim area, the Darling River traverses two surface water management areas constituted under NSW water legislation (*Water Management Act 2000* s 11). These are the upstream Western Water Management Area, in which the Barwon-Darling River runs mostly free of dams, and the downstream Lower Darling Water Management Area, which is affected and controlled by numerous dams and other water regulation infrastructure (Figure 1). The Menindee Lakes, which are approximately in the middle of Barkandji country and are particularly rich with Aboriginal cultural heritage [38,48], separate these two water management areas. The Lower Darling provides much-needed water resources to southern NSW water users and to the downstream States of

Victoria and South Australia, but it receives low rainfall and runoff, and is reliant on declining flows from the Barwon-Darling and other upstream tributaries [49].

In addition to suffering from decades of colonial control and exclusion, as well as ongoing intergenerational impacts stemming from these injustices, Barkandji must today contend with the pressure exerted by competition from other water users and dramatically declining river health. Indeed, NSW arguably has the most over-allocated water systems in Australia [11]. Tan and Jackson [11] suggest that embargoes on issuing new licences in NSW from as early as 1976 precluded substantial Aboriginal access to water. Moreover, the management of the Darling River and its upstream intersecting rivers, has been subject to ongoing controversy particularly over the past 20 years during which time the future of cotton production has been debated. Public discussion about water use in this region reached a pinnacle in July 2017, when a national investigative television program alleged that upstream cotton irrigators had been stealing billions of dollars of water [50]. Numerous government and criminal investigations into these matters are now underway (for example [51,52]).

To consider Barkandji experiences as a case study for this article, we rely on a mixture of sources for evidence. We draw on relevant government policy, legislative, and public inquiry documents pertaining to water regimes, native title law, and their overlap, at the NSW and Australian jurisdictional levels. We also draw from 17 interviews with 21 individuals, including Barkandji Traditional Owners, their legal representatives, and former NSW Government water agency employees, all of which were conducted by the lead author between February 2017 and January 2018. Interviews were conducted in accordance with the human ethics guidelines of Griffith University (GU Ref Nos: OTH/02/15/HREC and 2016/387). Barkandji Traditional Owner interviewees were identified through a snowballing strategy [53] beginning with the Directors of the Prescribed Body Corporate (PBC), which is the corporation set up under the NTA to hold and manage native title in trust for all Barkandji native title holders. Interviewees are named throughout this article in line with their preferences specified during consent procedures. We note that Barkandji language does not have distinct 'p' and 'b' sounds, nor 't' and 'd' sounds. For this reason, Barkandji words—including the word 'Barkandji'—can have many spellings [38]. In this article, we generally take the spelling 'Barkandji' as used in legal native title processes. We are also sensitive to other spellings from sources and as requested by interviewees.

4. Barkandji's Water Rights Struggle

The *Barka* is of great significance to the Barkandji People in interconnected cosmological and material ways. Central to Barkandji culture, spirituality, and teachings, the *Barka* is home to the *Ngatji* (Rainbow Serpent), who created the lands and the rivers. The Barkandji are responsible for the *Ngatji*'s health and wellbeing, although they find this increasingly outside of their control under contemporary water governance arrangements [45,54]. Barkandji Traditional Owners' physical, mental and social health is linked to the health of the River [43,45], with many Barkandji interviewees convinced that improved river health leads to lower occurrences of crime (see also [44]). As Barkandji PBC Director A phrased it, '*Without this water, we will never survive. We will be all* 'bukali' ... '*Bukali*' *means we'll die!*' (7 February 2017). While the Barkandji People's ways of thinking about and experiencing the Darling River have changed over time, partly in response to the displacement of their People and alterations to their country brought about by European occupation, the River has consistently remained central to their cultural identity [43,45].

Issues of growing water scarcity, compromised Darling River health, and the serious resulting impacts for the Barkandji Peoples, their landscapes, and other Darling River communities has motivated some Barkandji to take action. In 2016, water concerns sparked two Barkandji-led protests. Dissatisfied with the declining health of the river and lack of sufficient government response to the critical situation, these organized protests aimed to raise awareness with governments, politicians, and the broader Australian public [55,56]. The subsequent lack of government action in response to the protests was disappointing for Barkandji protestors (Barkandji Person C, 17 February 2017).

To take action and seek recognition, the Barkandji People, like many other Traditional Owners, 'have to argue for their rights and responsibilities to be recognized within the introduced European systems of law and governance' [57] (p. 182). We now turn to consider the four key regimes that determine the extent to which Barkandji's legal rights to water are recognized.

4.1. National Water Policy

Australia's National Water Initiative (NWI), an intergovernmental agreement negotiated by the Federal, State and Territory Governments in 2004 [6], established a nationally consistent approach to water reform, including a water entitlements framework and mandatory water planning. The NWI has been described as the most significant change in water policy since Australia's Federation in 1901 [58]. Building on previous national water reform commitments, the NWI called for clear entitlements to water, trade in water entitlements, transparent statutory-based water planning and environmentally sustainable management of water.

Although Aboriginal peoples played no part in its negotiation [41], the NWI represents the first national attempt to recognize Indigenous specific water rights and interests in policy [59]. By providing the impetus for Indigenous water needs to be recognized and accommodated by State and Territory water access entitlements and planning frameworks, the NWI therefore also set the scene for more significant recognition of Indigenous water rights opportunities across Australia's water governance mechanisms. Specifically, the NWI calls for Indigenous access to be achieved through:

- including Indigenous representation in water planning, wherever possible;
- incorporating Indigenous social, spiritual and customary objectives and strategies for achieving these objectives, wherever they can be developed;
- taking account of the possible existence of native title rights to water in the catchment or aquifer area;
- potentially allocating water to native title holders; and
- accounting for any water allocated to native title holders for traditional cultural purposes [6] (cls 52–54).

The NWI guidelines stipulate that water plans should immediately include the consideration of Indigenous water uses [59].

The emphasis these clauses place on 'traditional cultural purposes' have been criticized for inadequately including Aboriginal peoples' understandings, uses and relationships with water, and for precluding economic development options [59]. Tan and Jackson [11] additionally argue the NWI's goals are prejudiced by delay and difficulties in native title determinations, and that a low priority is given to Aboriginal needs in over-allocated catchments. Alongside these criticisms, reviews have consistently commented on the poor implementation of these actions [11,60–62]. The National Water Commission's 2014 review [62] found that after more than ten years there had been 'no substantial increase in water allocation for Indigenous purposes—social, economic or cultural' (p. 5). The most recent assessment of water reform progress notes some improvement in processes of consultation and engagement, but still calls for governments to better identify Indigenous objectives in water planning frameworks [63].

As the NWI remains the current instrument guiding legislative reform and water management across Australia, State and Territory Governments are required to comply with the aforementioned principles. However, no penalties are imposed on State Governments for poor or non-compliance; therefore, there is little incentive to really drive change that meaningfully recognizes Aboriginal water rights [11,64]. Regardless, the NWI sets the context in which the Barkandji Peoples—and indeed all Aboriginal peoples—can seek to have their legal rights to water recognized. Before considering the water legislation and water sharing planning regimes specific to the Barkandji in NSW, we first consider the native title framework and the associated water rights opportunities it offers within the context of the Barkandji's native title claim.

4.2. Native Title Law and the Barkandji Native Title Claim

The *Native Title Act 1993* (Cth) (NTA hereafter) is the Australian legislative response to the landmark *Mabo* High Court decision of 1992 that rejected the colonial falsity that Australia was *terra nullius* (meaning 'land belonging to no one') at the time of European occupation. The *Mabo* decision and NTA offered hope to those seeking recognition of the existence of two tenure systems: the introduced colonial system, from which land titles and water rights regimes flow, and the pre-existing and oldest surviving system of land tenure in the world, from which native title rights derive [65].

Native title is defined by the NTA as the communal, group or individual rights and interests of Aboriginal or Torres Strait Islander peoples in relation to land and waters (s 223). To have native title rights and interests in relation to a particular area of land (and waters) formally recognized, Traditional Owners register a native title claim that then goes through what is called a native title determination (NTA s 225). This is where the court 'determines' whether or not native title exists on a case-by-case basis, and if it does, then specifies the nature and extent of the native title rights and interests (as well as any other non-native title interests in the determination area). This recognition occurs after either litigation or the conclusion of an agreement between the native title claimants, relevant government parties and others with interests in that particular area, which is called a 'consent determination'. These processes are complex and costly, and can take many years to prove and settle [66]. Some or all native title rights and interests in a given claim may be found to have been extinguished in whole or part by past valid government acts (as defined in the NTA). The extinguishment of native title is permanent and cannot be revived even if the act(s) that caused extinguishment cease to have effect (NTA s 237A). Native title rights and interests vary from rights of exclusive possession in land to minimal rights of access for limited purposes [67].

As at January 2018, 375 litigated and consent determinations have been made across Australia, while just under 300 claims await determination [68]. The Barkandji's 2015 successful determination was only the sixth of its kind in NSW since the introduction of the NTA [47]. Settling native title claims in NSW is noted to be a slow process compared to other State and Territory jurisdictions [69].

The trajectory of native title law and its practical outcomes has proven disappointing to Aboriginal peoples and their advocates [35,65]. The regime's emphasis on pre-European 'traditional' rights and interests and the requirement for claimants to prove the ongoing and unbroken existence of these rights in a demanding evidentiary process has created substantial hurdles, as well as precluded the evolution of native title rights [29,70]. In addition to protecting these Aboriginal rights and customs, the NTA also protects and upholds existing land and water title holders, validates past actions which may impair native title interests, and regulates future ones [35]. Thus, in intensively colonized areas such as the State of NSW, the difficulty of proving ongoing and unbroken continuity to place required through legal native title determination processes is heightened [1,71]. Furthermore, State Governments have been slow to accommodate and protect native title rights, if at all, in their land and water management legislation and policies [35,64,72]. As a result, there are concerns about the effectiveness of native title in protecting Traditional Owners' abilities to exercise their inherent rights, enforce their traditional laws and governance institutions, and control their resources, including water [64].

Limitations in the treatment and recognition of Aboriginal water rights by the native title regime have been well documented [3,11,12,29]. O'Donnell [12] argues there are two propositions that are clear in relation to Aboriginal native title rights to water. The first is that native title does not include ownership of natural waters on the assumption that the common law's position is that water in its natural state is not amenable to ownership. The second proposition is that where native title can be proven to exist, it generally includes rights to take and use water for only personal, social, domestic and cultural purposes. It can include a:

- right to teach the physical and spiritual attributes of places and areas of importance on or in the land and waters;

- right to have access to, maintain and protect places and areas of importance on or in the land and waters; and
- right of access to take water for those purposes.

These rights have been found to apply to flowing, surface and subterranean waters. To date, a native title right to take and use water for commercial purposes has not been recognized [12], an interpretation that, as many commentators observe, constrains native title holders' water access and utility [3,29,64,72]. While the possibility for economic uses and benefits of native title rights to natural resources, including water, has emerged in recent years, including as a recommended area for legislative reform [66], this has not yet eventuated.

Similar to many native title determinations [12], the Barkandji have a right to take water for drinking and domestic uses. Their water use right is specified as being:

> for personal, domestic and communal purposes (including cultural purposes and for watering native animals, cattle and other stock, and watering gardens not exceeding 2 hectares), but not extending to a right to control the use and flow of the water in any rivers or lakes which flow through or past or are situated within the land of two or more occupiers. [46] (para 6)

The term 'cultural purposes' in this water use right was also prescribed in the Barkandji's court determination to include the purposes of performing activities of a cultural nature that 'involve the use of insubstantial quantities of water' such as 'cleansing ceremonies'; 'the preparation of food or bush medicines'; and 'activities involving the teaching of native title holders about traditional laws, customs and practices' to list just a few (see [46]). Ancillary rights and interests that indirectly relate to water were also recognized, including (though not limited to) the right to hunt and fish; the right to take and use natural resources (other than water); the right to engage in cultural activities; and the right to have access to, to maintain and to protect from physical harm sites and places of importance or significance under traditional laws and customs [46] (para 6).

These and the Barkandji's other native title rights and interests can be enjoyed in specified areas alongside others' existing (non-native title) rights. Notably, a 400 km stretch of the Darling River, and several water courses and lagoons in the south of the claim are part of these specified areas. Other Barkandji river country was excluded from the Barkandji Traditional Owners #8 claim [46] (para 13), such as the lower reaches of the Darling River (see Figure 1), because at the time this 1997 application was made, another native title application covered these areas and the NTA regime prohibits overlapping claims (F. Russo, 27 June 2017).

The NTA also makes specific provision in relation to native title rights to water by:

- confirming Crown or government rights to the use, control and regulation or management of water;
- validating any water management legislation that was enacted between 31 October 1975 and 1 July 1993 (the period between the introduction of the *Racial Discrimination Act 1975* (Cth) and the NTA);
- confirming 'existing' public access to and enjoyment of waterways, beds and banks or foreshores of waterways, coastal waters and beaches where native title exists;
- preserving certain native title non-commercial activities in relation to water from some types of government regulation in Section 211 (meaning no licences are required); and
- providing a future act regime to regulate how government and third parties can affect or impact native title rights to water including procedural and compensation rights in Section 24HA [73].

Under these future acts measures, any registered native title claimants and native title holders, like the Barkandji, have the right to be notified prior to the grant of any water management or regulation related lease, licence, permit or authority that might affect their land or waters. Native title holders and claimants are given the opportunity to comment on—though not object to or prevent—any proposed action/s. Significantly this does not apply in the making, amendment or repeal of water management or regulation legislation. Native title holders also have the right to compensation where these acts affect native title. While compensation may take the form of financial payments and/or include opportunities for employment, training and education, or cultural site protection, rehabilitation or monitoring, claiming and payment of compensation is still an emerging aspect of the NTA regime [35,74]. In any combination, these weak measures can constrain Aboriginal peoples' participation in water resource management in that they create 'legal certainty for States and third parties at the expense of native title' [11] (p. 140).

Indigenous Land Use Agreements (ILUAs) offer a potential vehicle for addressing and leveraging these water-related procedural and compensation rights [11,72,74,75], but assessing their effectiveness is difficult as ILUAs are generally reached in-confidence [75]. At the time of writing, Barkandji native title holders are in the early stages of ILUA negotiations with the NSW Government, which may at some stage include measures to address their water rights and management concerns.

As the previous section showed, national water policy includes specific provisions relating not only to native title holders but also, the possible existence of native title [6] (cls 52–54). We now turn to the NSW State-based water legislative regime, which is expected to comply with this national policy, to see how these native title specific requirements have been implemented.

4.3. NSW Water Legislation

Bringing about significant change to the previous State water legislation, the NSW *Water Management Act 2000* (WMA hereafter) is a legislative response to over twenty years of national water reform [76,77]. The WMA notably includes a broad objective to 'recognise and foster the significant social and economic benefits … to the Aboriginal people in relation to their spiritual, social, customary and economic use of land and water' (s 3). To meet this objective, several provisions are included in the legislation, some of which have been described as relatively more progressive and advanced compared to other jurisdictions [61,62], though cautiously so by some [3,11,78].

First, the WMA provides that the Minister can establish multi-stakeholder catchment-level water management committees of at least 12 but no more than 20 members, of whom at least two are Aboriginal persons (s 13). Across Australian jurisdictions, NSW is the only State that stipulates provisions for Aboriginal representation in this manner [11]. This measure was an improvement on previous legislation that barely recognized Aboriginal water rights [76] and tended to limit consultations to existing water licence holders only. However, this model of consultation nonetheless raises numerous problems and challenges unique to Aboriginal contexts. The prevailing power imbalances between Aboriginal and non-Aboriginal peoples broadly within Australia and in water resource management specifically, as well as the related comparatively recent recognition of Aboriginal peoples' water rights and interests in Australian water policy, largely underpin these Indigenous-specific obstacles, as does Aboriginal peoples' ontological foundations and sense of obligation to country and community [10]. For example, one or two individuals face difficulties in representing numerous Aboriginal Nations with rights and interests within a legislated water management area [79]. Without adequate support, Aboriginal representatives may find it difficult to fulfil their responsibilities and obligations to other members of their Nation, particularly amongst widely dispersed populations like those of the Barkandji. The technical and legal complexity of many of the hydrological issues and entitlement frameworks addressed by advisory groups may also inhibit effective participation [1,10,79].

While these legislative provisions for water management committees including Aboriginal representation remain in the WMA, they are no longer used (B. Moggridge, *pers comm*, 22 December 2017). In fact, these committees were only used until 2004, four years after the introduction of the Act, when water planning policy and practice changed [78] to 'mainly involve *bureaucratic* coordination and bargaining' within government departments [76] (p. 80, emphasis in original). This practice is compliant with the WMA legislation given it is the Minister's discretion to use these water catchment committees.

More recently, in 2016 and 2017 as part of 'new governance arrangements' to inform future water resource planning in parts of NSW, Stakeholder Advisory Panels (SAPs) have been established, each with 14 or 15 members [80,81]. This change was part of a multijurisdictional coordinated water planning exercise for the Murray-Darling Basin. Explicit discussion of this overlapping water regime is beyond the scope of this article. The provisions of the NSW Government's WMA including WSPs and water licensing processes operate independently of this overlapping regime [82]. Aboriginal input into water planning for these areas now occurs via individual representation in such Panels [83]. Regardless of how many First Nations' lands and waters each water planning area traverses, there are provisions for one Aboriginal community representative member on each surface water SAP [80], and two on the single State-wide groundwater SAP [81]. This model is akin to the individual representation of the WMA's water management committees and thus suffers from deficiencies similar to those discussed above. Management of water resources within Barkandji country will be informed by at least three SAPs, two of which currently have one Barkandji representative each.

Second, the WMA and its supporting regulation introduced Aboriginal specific licences for the first time [77]. Such licences include (a) cultural access licences; (b) Aboriginal community development licences; and (c) Aboriginal environment licences. While the availability of these specific purpose licences have been celebrated [61,62], they are not all available in all areas of NSW. For example, while the cultural use licences are available in all surface water and groundwater management areas, the community development licences are only available in catchments where water extraction is not yet over allocated, namely in coastal water management areas [77]. The Aboriginal environment licences are for supplementary water, the name given to periods of high river flows, and to date, are only available in relation to the Barwon-Darling River [84] (cl 50). All of these Aboriginal specific licences are conditioned with limits to volumetric entitlements, use options, and restrict or outright prohibit trade, and the actual uptake has been low for a number of possible reasons [3,11]. Barkandji native title holders have not applied for any Aboriginal specific purpose licences.

Third, NSW water legislation provides for native title rights. NSW is one of very few jurisdictions to have done so despite the NWI expectation that native title water rights be accounted for by all Australian State and Territory water regimes [72]. Under Section 55 of the WMA, water required to exercise native title rights are reserved as 'Basic Landholder Rights' and so are afforded the same priority as domestic and stock rights of riparian land owners or occupiers. Accounting for native title water rights as Basic Landholder Rights notionally positions them in the highest category of water rights as these water requirements must be met first, prior to any other consumptive water uses, even in extreme drought conditions [11,72]. Duff [72] notes that this outcome is a consequence of recognition of the native title holders' rights in relation to land; it is not dependent on any native title rights specifically in or to water. While Macpherson [29] notes that Aboriginal water rights have been 'shoehorned' to fit within this framework, perhaps as a means to reduce conflict between Aboriginal peoples and other water uses, this provision is nonetheless the mechanism for accounting for and recognizing water-dependent native title rights in NSW. Perhaps most fundamentally though, an Aboriginal group must first have overcome the hurdles of the Commonwealth native title legislation and have their rights determined for these, albeit limited, native title provisions to apply [1].

Commentators have raised three other criticisms about these NSW native title allocations. The first is that Basic Landholder Rights—including domestic and stock rights and native title rights—do not require a water access licence to take and use water (WMA s 55). As mentioned earlier though, the NTA already provides licence exemptions for native title holders, and so this specific legislative allowance is insignificant [72]. Second, as Behrendt and Thompson [1] point out, 'an entitlement to extract water does not ensure that there is any water to extract or that the water is of consumable quality' (p. 97). Thirdly, the majority of Water Sharing Plans (the regulatory water management instrument in which Basic Landholder Rights are quantified and protected—discussed below) have zero allocations for native title, rendering any apparent priority for Aboriginal water allocations as illusory [11].

The legislative provisions discussed here are implemented and operationalized through the NSW water allocation planning regime (also referred to as water sharing), which we now consider.

4.4. NSW Water Allocation Planning Regime

Water sharing, a key element of the National Water Initiative, is considered to be a fundamental tool for achieving sustainable water use. In NSW under the WMA, water sharing has been regulated through the progressive development of over 80 Water Sharing Plans (WSPs) for surface water and groundwater systems [85]. WSPs are regulatory instruments that contain enforced 'rules for sharing water between different types of water use such as town supply, rural domestic supply, stock watering, industry and irrigation and ensures that water is provided for the health of the system' [86]. As a regulation under the WMA, any breaches to WSPs attract possible litigation and/or monetary penalties (WMA s 336). WSPs are generally in operation for ten years, after which they are replaced or extended, and may be suspended in times of severe water shortages (WMA s 49(a)).

Under the NSW State water legislation, WSPs can be made via two processes, either as 'Management Plans' with the involvement of the above mentioned multi-stakeholder water catchment committees (WMA Part 3), or as 'Minister's Plans' without their involvement (WMA s 50). The initial 31 WSPs—which covered about 80% of water extracted within NSW—were prepared through water management committees. Once prepared, however, the majority were over-ridden and re-drafted as Minister's Plans by the NSW Government [78]. This over-ruling attracted much frustration [76] and litigation against the Minister [78]. Minister's Plans are not a secondary or subordinate form of WSP as determined in the aforementioned litigation cases (see particularly [87] at para 35), and as such, both formats comply with the NWI. Our review of the WSPs currently in operation reveals that all are Minister's Plans. Significantly, when preparing this type of Plan, the Minister retains discretion regarding stakeholder engagement and which notification provisions to adopt (WMA s 50(2A)) (see also [78,79]). While Aboriginal people contributed to the development of Minister's Plans through a dedicated Aboriginal unit within the NSW water agency (which existed from 2012 to 2016) [88], overall, the reliance on Minister's Plans limits opportunities for Aboriginal peoples, as well as the general public, to provide sustained and comprehensive input. This is compared to planning processes that directly involve multiple stakeholders and contain opportunities to deliberate over water-use scenarios and impacts while considering trade-offs amongst competing uses [89].

Native title water-related rights in NSW, established under the three regimes discussed above—national native title legislation, national water policy and NSW water legislation—are brought together and operationalized in WSPs as Basic Landholder Rights. All WSPs must deal with certain matters, regardless of Plan development method, of which protecting Basic Landholder Rights is one (WMA s 20(1)). WSPs identify and specify Basic Landholder Rights, including domestic and stock rights and native title rights, so they can be satisfied first before other water needs in each water management area (WMA s 20(1)). One WSP in coastal north-east NSW, for example, specifies that native title holders, being the Yaegl People, the Bandjalang People, and the Githabul People, 'are entitled to take [water] pursuant to their native title rights', without specifying volumetric limits [90] (cl 20). In contrast, another coastal WSP reserves a volumetric amount—26.6 megaliters per year—for native title requirements [91] (cl 21). As mentioned, however, the majority of WSPs have zero allocations to satisfy native title rights [11].

Although all WSPs can be updated to give effect to successful native title determinations that are handed down through specific amendment allowances, Behrendt and Thompson [1] call this a 'don't worry about it until it arises' approach, which they regard as 'far from satisfactory' (p. 104). Exactly how the accommodation and protection of native title water rights are incorporated into WSPs—either during initial development, or once a WSP has commenced and a native title claim is determined—is difficult to ascertain. Policy and government document requests offered negligible clarification. Only three former employees could assist, explaining that the procedure to incorporate native title depends on NSW Government employees first knowing a native title claim exists and the outcome of the determination, and second, notifying the appropriate water planner for the relevant management area so that it can be accommodated in the Plan. In other words, it is an ad hoc, manual process that one former official acknowledged *'didn't work very well'* (L. Betterridge, 16 June 2017). This is evidenced by the fact that, where native title rights have been included in WSPs, this only occurred because legal representatives of native title claimants or holders raised the matter with the NSW Government and not through any government-driven process [11,92]. There is, therefore, little transparency or accountability, which complicates monitoring and reviewing, or indeed contesting, the NSW Government's approach to the protection of native title.

In the Barkandji's case, five WSPs overlap with the areas where their native title rights and interests have been recognized (including a 400 km stretch of the Darling River). Four of these overlapping WSPs commenced at various times throughout 2012 [84,93–95], several years prior to Barkandji's native title determination of 16 June 2015, but nonetheless, during the time their claim was being negotiated by the NSW Government. The fifth WSP, the *WSP for the New South Wales Murray and Lower Darling Regulated Rivers Water Sources 2016* ('2016 Murray and Lower Darling WSP' hereafter), commenced on 1 July 2016 [96], more than one year *after* the Barkandji's native title determination was handed down. Yet, as of January 2018, more than 2.5 years after the Barkandji's native title rights were affirmed by the court, all five WSPs state: 'At the commencement of this Plan, *there are no native title rights in these water sources.* Therefore the *water requirements for native title rights are 0 ML/year'* [84,93–96] (emphasis added). Despite the existence of amendment mechanisms designed to reflect changes to native title during their operation and the Barkandji's best efforts to bring this error to the NSW Government's attention (see further discussion below), none the five WSPs has been updated, and so they violate the native title provisions in both the NSW Government's WMA and national water policy.

In particular, we wish to focus briefly on the 2016 Murray and Lower Darling WSP which, again, was introduced one year after the Barkandji's native title determination. When introduced (and still at the time of writing, 18 months later), it stated incorrectly that there are 'no native title rights' in the area [96] (cl 19). Arguably, this Plan was, then, invalid when written and gazetted. The initial Murray and Lower Darling WSP commenced in July 2004, with the 2016 Plan only a 'replacement' [97]. Some 'minimal changes' were made though, including 'updating share components and Basic Landholder Rights estimates' [97], which, as established above, should encompasses native title water requirements but did not. While there was 'no formal public consultation process undertaken', the NSW Government claims to have 'consulted with key stakeholder groups to seek feedback on changes required in the plan' [97]. It is possible that the flexibility and selectivity of this stakeholder engagement and consultation method stems from the discretion allowed for through Minister's Plans under regulations.

Interviews verified that the Barkandji People were not consulted about these changes to the replacement 2016 Murray and Lower Darling WSP despite the occurrence of this active, though limited, stakeholder engagement process. Consultation activities with 'key stakeholders' listed online reveals no specific mention of the Barkandji People either [98]. Ultimately, the Barkandji only learned about the 2016 Murray and Lower Darling WSP and its failure to recognize and accommodate their established rights once it was implemented. Failing to meaningfully engage with the Barkandji and accommodate and protect their native title rights, as they are entitled to, raises serious questions about the validity of the updated estimates of Basic Landholder Rights and undermines the claimed 'consistency with the current legislative framework' [97] of the updated WSP.

The practical implications from not recognizing or accommodating Barkandji native title rights in these WSPs have not yet been tested. It would seem from secondary analyses of native title legislation [72] that failure to recognize native title rights does not prevent the Barkandji People from exercising their rights over water because Commonwealth laws (i.e., the NTA) prevail over State and Territory laws (i.e., the NSW's WMA). The WSPs are, however, regulatory instruments, and non-compliance can attract fines and prosecution. It could be possible that Barkandji People, in exercising their native title rights to take and use water that is not allocated for their use in the WSP, may be perceived by compliance officers as breaching the regulation. While such a situation may seem surprising or unlikely, this has, in fact, occurred in relation to cultural fishing rights in South Australia (see [99]), and to an extent in NSW (see [100]), and cannot be ruled out as a possibility in relation to water. As is evident from these cultural fishing cases, such prosecution certainly can be challenged successfully using a native title defense, but is a time- and resource-intensive exercise. The need to defend a right that has already been recognized appears to be superfluous.

Perhaps, though, the biggest impact these exclusions from the WSPs has had is the failure to protect water and maintain sustainable water levels that support Barkandji's enjoyment and exercise of their other water-related native title rights and interests (detailed earlier in Section 4.2). This includes, for example, the right to maintain and protect sites and places of importance or significance from physical harm, which could require significant water volumes [72]. Water needed for this kind of activity is likely to be unavailable due to complete or over-allocation of water resources in the water management areas, and would thus require the acquisition of water from existing water users, and potentially trigger compensation claims (Aboriginal Water Planning Specialist, 22 January 2018). Barkandji themselves do not have the power to reallocate this water—this lies with the State of NSW. It seems the NSW Government faces a most difficult challenge in ensuring native title rights and interests are accommodated and protected in accordance with water regimes discussed here when all available water is already allocated to existing water users (Aboriginal Water Planning Specialist, 22 January 2018).

4.5. Barkandji Experiences and Perspectives

The Barkandji Traditional Owners have made several attempts to challenge this disregard by the NSW Government. NTSCORP Limited ('NTSCORP' hereafter), the Barkandji Peoples' legal representatives, have been assisting in these attempts, which initially focused only on the 2016 Murray and Lower Darling WSP, but expanded to include other WSPs following further inquiries and research [101]. In addition to representing the Barkandji, NTSCORP has statutory responsibilities under Part 11 of the NTA to protect Traditional Owners' native title rights and interests across the State of NSW (and the Australian Capital Territory), including assisting, servicing and representing native title claimants and holders. The Barkandji first approached the NSW Government about the issue via email in July 2016 soon after the 2016 Murray and Lower Darling WSP commenced. The details of the Barkandji's native title rights and interests were set out in this correspondence, including the waters within their determination, and their overlap with this WSP (F. Russo, 27 June 2017). The NSW Government advised the matter would be investigated and the WSP updated if needed. As of November 2017, no further formal response had been received.

The issue was again raised later in 2016 as part of a NSW Parliament Inquiry into water supply for rural and regional NSW. Among other matters, NTSCORP's formal submission highlighted the cultural insensitivity and misleading nature of NSW's water sharing regimes that do not respond to legal recognition of native title rights. NTSCORP, from the position of representing the Barkandji People and other Traditional Owners across NSW, noted,

> it is extremely distressing for Traditional Owners when the NSW Office of Water publishes Water Sharing Plan[s] that do not acknowledge successful native title determinations or pending claims of native title rights and interests within the area covered by the Plan. [102] (p. 15)

As part of this inquiry, Barkandji native title holders were invited to present at a regional public hearing on these matters as well. The above-mentioned frustrations are captured in Barkandji PBC Director Mr Badger Bates' opening statement: 'Our Barkandji native title gave us recognition but not much else ... we hoped that our recognised native title will give us the right to manage our river for future generations' [92] (p. 2). This inquiry is ongoing and due to be reported on by 30 March 2018. Again, no feedback or advice from the NSW Government has been received in response to these matters. (We have also made our own inquiries with the NSW Government on this matter. Our original inquiry was submitted on 21 November 2016, to which we received a response suggesting, incorrectly, the Barkandji's native title determination 'excluded the waterways and therefore they [the Barkandji native title rights] are not reflected in the WSP' on 28 February 2017. Clarification on this response was sought on 31 July 2017, and as at January 2018 we are awaiting a response.)

According to NTSCORP, this lack of recognition in the water allocation planning regime is affecting the exercise of Barkandji's native title procedural rights concerning water regulation and management (under s 24HA of NTA). The Barkandji seek to communicate their opposition to further water extractions via their right to comment on proposed actions that may affect their native title rights to water, including the granting of new, and extensions to existing, water use and extraction licences. While acknowledging this procedural right is not a right to object or veto, an NTSCORP representative noted that the relevant NSW Government department disregards the native title holders' comments and grants the water use licences (F. Russo, 27 June 2017). The NSW Government instead:

> ... pointed out that the method to deal with any such objections is through the public submission process for the Water Sharing Plans. So it's a circular argument—the Water Sharing Plans are already in place, and they haven't been updated in light of the determination of native title. Yet our clients' rights are meant to be factored in through those Plans rather than through objections to individual licences being granted. (F. Russo, 27 June 2017)

The Barkandji have also commented on water extraction approvals for the construction and use of infrastructure like pumps, bores and jetties, requesting that conditions be applied to these licences to protect and monitor Aboriginal cultural heritage in ways that align with NSW's existing cultural heritage management guidelines. The NSW Government—a different department—has been more agreeable to these requests, imposing the requested conditions on new licences, including *'even in instances where we haven't responded to a particular notice, they're now starting to include conditions to ensure that licence holders are aware of their responsibilities to protect Aboriginal cultural heritage'* (F. Russo, 27 June 2017). The discretionary and ad hoc nature of government responses to Aboriginal peoples' (and specifically, native title) water and water-related rights, which is at least partially underpinned by a lack of constraints to guide these responses, is indicative of asymmetrical power relations.

Interviews with Barkandji spokespeople and Traditional Owners reveal that they regard these events and ongoing struggles to have their native title rights appropriately recognized and accommodated as frustrating, disrespectful and insensitive. Barkandji native title holders feel that while they are expected to respond to various government requirements they receive little to no government response to their own requests. They perceive there has been little attempt by governments to provide meaningful opportunities for their engagement, demonstrating the limitations of these water governance processes to generate opportunities for mutual recognition, collaboration and partnership in governance, no matter how difficult satisfying such aspirational goals may be in a settler state. As Barkandji PBC Director B suggested, *'We never got no contacts or nothing from the government about the river! Didn't even come out and sit down with us'* (7 February 2017). Importantly, this frustration extends beyond water allocation planning discussions and encompasses a wide range of other Barkandji water-related concerns.

Many Barkandji interviewees see continued government control as a central obstacle to the appropriate recognition of their legal water rights. As one interviewee passionately described, *'At the end of the day, whatever we want put in place, the governments have already got their agenda'* (Barkandji Person B, 15 February 2017). By extension, some express disappointment in the level of respect they have been shown, as native title holders. These overlapping issues were highlighted by Barkandji PBC Director C:

> Now that we've got our native title claim, they should be negotiating with us and consulting with us. We are the experts and the land managers ... [But] we have to put everything up to them. They sit in their offices and make decisions from there. We want them to come out here and consult with us. (7 February 2017)

These observations of power asymmetries are not limited to water, but spread across natural resource governance. As one interviewee noted, *'The government's always got an ulterior motive with Aboriginal people. They say one thing and do another'* (Barkandji Person C, 17 February 2017). Another noted, *'The water seems to be a different issue to the land issues, and they [government] want to keep it that way because it's about control'* (Paakindji Man, 9 February 2017). A parallel issue is the influence and seeming dominance of other (usually agricultural) interests over those of the Barkandji, particularly the interests of upstream cotton growers and foreign investors. Barkandji PBC Director E went as far as to say they see their native title determination as a *'flimsy agreement'* that is *'not real good'* due to the little power it affords their community when negotiating with other interests, including governments and other water users (10 February 2017).

Despite having been recognized as the native title claimants for nearly 18 years, and then since June 2015 the native title holders, the Barkandji feel they face an unfair and ongoing struggle for recognition and representation in water resource planning processes affecting their country. We now turn to more explicitly reflect on the Barkandji's experience in light of the insights elicited from the politics of recognition literature, and to consider wider implications for the recognition of Aboriginal water rights within Australia's water regimes.

5. Discussion

Ivison [14] argues that to dislodge the dominant focus on recognition, institutions must be open to processes of critical reflection and challenge regarding the outcomes they produce. Here, we illuminate some of the unjust outcomes resulting from attempts by the state to recognize Aboriginal peoples and their water rights, or make them legible and visible, through Australia's water regimes; outcomes which we suggest serve to defer and diffuse more radical Aboriginal struggles and challenges of state sovereignty, in that the power and control remains in the hands of the state [18]. For the purposes of analysis, we have broadly grouped these outcomes into two forms; 'misrecognition' and 'non-recognition', which are discussed in turn. Notably, while these outcomes are likely unintentional and possibly a result of failed bureaucratic processes, as we will discuss, determining or proving intentionality is a demanding task that exceeds the evidence obtained during the course of this study.

Building on James C. Scott's [16] work on legibility, we argue that forms of misrecognition can be understood as a process that renders Aboriginal peoples more governable: to frame any rights within existing, colonial systems of management, governance, and engagement so as to support, rather than challenge or subvert those systems. We identify two ways in which Aboriginal peoples are misrecognized through, or as a result of, recognition processes in Australia's settler-colonial water regimes. The first is what we term stakeholder-based misrecognition, which has the effect of rendering Aboriginal peoples as no different to other water users, with 'equal rights granted to all groups in the multicultural nation' [18] (p. 57). This politico-legal process has effectively allowed governments to allocate 'water entitlements with little regard or knowledge of Indigenous interests' [3] (p. 109), a course of action which many Aboriginal peoples believe has amplified their historical and contemporary inequities.

Consequences associated with stakeholder-based misrecognition include the possibility that the state and other actors may overlook custodial rights and responsibilities or special rights of access and management arising from international Indigenous rights norms. This then, has the effect of eliding and hampering the legitimacy afforded by others to Indigenous peoples' rights and responsibilities. Questions of sovereignty, justice, and reparations are wholly sidestepped by emphasizing the liberal conception of multicultural 'inclusion', and thus there is no platform from which to challenge the (colonial) logic of planning and management structures. Overall, stakeholder-based misrecognition has the effect that all stakeholders are positioned as, theoretically at least, 'equal', and this can weaken the political merit of any particular claims flowing from Indigeneity. It also has the effect of obscuring any structural and power imbalances that may exist between and across parties. Additionally, such an approach also effectively situates the state as arbiter of different sets of (often competing) interests, making decisions from a position of constructed neutrality after weighing up input from many parties, obviating the state's own interest in particular outcomes, or affinities with particular stakeholders or sectors of society.

Over the past 15 or so years, an alternative type of misrecognition of Aboriginal peoples has resulted through water regime recognition processes. Under this alternative, Aboriginal peoples' rights to and interests in water are conceptualized as unique and different from other stakeholders', but only within the confines of (what the state mediates to be) tradition and culture. In this way, Indigenous knowledge—and indeed, values and rights—are shaped and must adhere to established and recognized forms of knowledge and representation, and in turn, remain on the periphery [23], or in a highly compartmentalized category of interest referred to in resource management discourse as 'cultural values' [103]. While this outcome from recognition strategies is in some ways progressive compared to the aforementioned stakeholder-based misrecognition outcome, and in part addresses Traditional Owners' demands for recognition, it has the effect of perpetuating a reified, narrow and essentialist conception of Aboriginal peoples [18] in ways akin to those seen in other environmental and land governance mechanisms, such as native title and land rights [3] and cultural heritage legislation [17,103]. Again, this process of flattening out complex ways of knowing, being and relating can be understood as a way to render a population knowable and legible—a way of making or keeping them as governable subjects of the state [16]. This misrecognition outcome turns formerly fluid, dynamic, and relational modes of identity and community formation into disciplined, fixed, simplistic and essentialized categories that existing forms of settler-colonial politico-legal processes are more equipped to deal with [16–18,21,33]. Thus, we refer to this outcome as an essentialist form of misrecognition.

Under Australia's current water regimes, the capacity of Aboriginal people to obtain and use water is greatly influenced by typologies of water use, which underpin water entitlements and use permits. This use-based interpretation and implementation that is central to Australia's water resource governance frameworks can be understood as another practice of legibility by the state. That is, it is a construction applied to water resources as a way to make water itself more knowable and governable. Attempting to translate their own diverse water needs, rights and objectives into the language of water regimes so that they may be recognized and accommodated, Aboriginal people have also generated proposals for water redistribution based on cultural-difference [3,13,42]. The concept of 'cultural flows', for example, has been proposed, which is defined as 'water entitlements that [would be] legally and beneficially owned by the Indigenous Nations of a sufficient and adequate quantity and quality to improve the spiritual, cultural, environmental, social and economic conditions of those Indigenous Nations' [104]. This complex cultural flows proposal, however, has been simplistically and restrictively translated by policymakers in essentialized ways to concern only what the state recognizes as 'cultural' water uses, which bureaucrats and policymakers have assumed will require insubstantial water volumes (for example, NSW Aboriginal specific licences, discussed in Section 4.3) (see also [3]).

The proscriptions underpinning essentialist misrecognition and the (restrained) rights it affords, ultimately curtails the emancipatory or transformative potential of recognition. As a consequence, those Aboriginal views or objectives that differ from what the state perceives and constructs as legitimate 'traditional cultural' practices are difficult to recognize or accommodate [18,103,105]. Moreover, some have recognized that entrenching reified constructions of Aboriginal peoples into governance structures in this way, even if unintended, can have the effect of marginalizing their voices [103,105,106]. These outcomes can severely limit the capacity of recognition to strengthen Aboriginal claims for self-determination and economic self-reliance, or the capacity to respond to other powerful stakeholders. Essentialist misrecognition does result in some recognition, but it does so in ways that fix, limit, and proscribe; it does not emancipate or transform and it remains mediated and determined by the settler state, making it incomplete [18].

From examining the Barkandji's water rights struggle case study, we can also see outcomes that we term non-recognition. Non-recognition describes situations where Aboriginal peoples and their rights go unrecognized—perhaps unintentionally—as a result of institutionalized and bureaucratic systems. The outcome of non-recognition forecloses opportunities to access and benefit from respect and possible acts of redistribution that could flow from due recognition. If misrecognition 'is a form of disrespect, and more strongly, denotes an absence of genuine mutual *esteem*' [14] (p. 14, emphasis in original), then outcomes of non-recognition are even more harmful and oppressive—regardless of intentionality.

Several possible and likely overlapping factors can help to explain this kind of failed recognition, as seen with the Barkandji's experience with the NSW water allocation regime (Section 4.4 above). There may be issues with inadequate system and notification processes across NSW Government departments during negotiations and once native title claims are determined. These failings may lead to gaps in, or delays to, departmental awareness and responses to successful determinations (see [107]). There may also be an issue of protecting and accommodating possible and actual native title rights specifically within the water-related bureaucratic agencies of the NSW Government. For instance, even though native title rights are positioned notionally as an upfront component to NSW's water allocation planning regime, these rights are only accommodated when detected through 'manual' and 'ad hoc' processes, based on imperfect governmental staff knowledge. Thus, the absence of an approach in NSW water planning and practice that systematically, consistently and thoroughly identifies and protects native title rights and interests and is tailored to each determination [1], likely contributes to non-recognition. It is important to identify that this disconnection between the legal arm of the state that recognizes native title rights, and the bureaucratic arm that ought to acknowledge and protect them, has the effect of limiting and undermining the (arguably, essentialist and already weak) recognition of Aboriginal peoples' water rights conferred through the native title regime [35].

If resources and expertise within government agencies are insufficient, this can also contribute to the ability of bureaucracies to grant due recognition. NSW Government water agencies and departments have faced political water cycle changes and undergone widespread restructures over the last several years, including the dismantling of the once 11-person Aboriginal-specific water unit [88]. Such changes have reduced the capacity of government to address these issues, according to Aboriginal people previously employed in that unit (B. Moggridge, 4 May 2017). Beyond individuals and teams of bureaucrats, though, political will from departmental leaders and governments is also needed to respond to complex and new issues of water sharing and allocation, particularly in areas where water resources are already completely or over-allocated. Meeting broader Aboriginal peoples' water needs will more than likely require the politically controversial commitment of reallocating water away from existing users [3]. There are also technical challenges, as Jackson and Morrison [59] note, in that 'there are substantial conceptual and technical difficulties facing water resource managers seeking to calculate and allocate water to meet these 'native title' needs,' (p. 30) irrespective of the water volumes that are actually available, and the political will for these changes. The failure of the NSW Government to respond to the Barkandji's requests could represent a deliberate strategy to withhold or deny due recognition of their rights, but it can just as equally be interpreted as the bureaucracy not knowing how to handle these complex issues, an indicator of the inadequacy of existing structures and systems, and/or a lack of commitment to prioritizing and addressing these issues.

Overall, though, by the virtue of the fact that settler states have claimed exclusive responsibility for overseeing the management and allocation of water resources, it similarly holds the responsibility to respond to and address these difficult challenges. This exemplifies the asymmetrical power relations in Australia's water governance between Aboriginal peoples and the colonial state. Under current water regimes, the Barkandji People hold neither the power to allocate themselves water, nor the power to force the state to protect and accommodate their rights. For this reason we argue that state failure to recognize and protect Barkandji native title rights in its water resource regimes challenges the legitimacy and justice of contemporary water governance in Australia. Furthermore, it also reveals the settler state's unjustified power to allow their institutions to continue to produce outcomes that benefit certain constituent actors while simultaneously and unfairly denying power and agency to Aboriginal peoples.

6. Conclusions

Ivison [14] advocates that disrupting the dominant focus on state-recognition requires greater attention to 'the ways in which current social and political arrangements manifest distinct forms of unjustified exercises of power' (p. 17). Heeding this call, we have critically examined the settler-colonial water regimes that are currently used by states to recognize Aboriginal peoples' water rights and control water in NSW. In doing so, we see that state attempts to recognize, or make legible, Aboriginal peoples and their water rights can produce outcomes that have the effect of actually misrecognizing Aboriginal peoples' claims to water either through overlooking and ignoring them (stakeholder-based misrecognition), or oversimplifying, stereotyping and restricting them (essentialist misrecognition). Intentionally or not, states may also outright fail to recognize any kind of claim, an outcome we have called non-recognition. It is concerning that non-recognition is possible following affirmation by the legal arm of the state, as is the experience of the Barkandji. Ultimately, the effect of state-based recognition perpetuates the status quo where existing water users hold the power to continue to enjoy and benefit from access to highly valuable water resources, while power and agency for Aboriginal peoples to do similarly remains obstructed. These manifestations of colonial power relations fail to generate genuine recognition and respect, and in turn undermine the legitimacy of state water regimes.

To conclude, we wish to reiterate several suggestions to enhance Indigenous water resources governance in Australia that have been recommended for more than a decade by both academic (for example, [10,11,59,64]) and government reviews (for example, [62,63]). Specifically, at the NSW State level, we argue that the NSW Government needs to instigate a more systematic network notification process to inform and engage all relevant departments and agencies—including water bureaucracies—during native title negotiations and once determinations are passed. Complementing this, specifically within water planning and practice, there is a need for greater commitment and obligation 'to identify 'and engage with' all beneficiaries and interests affected by planning up front' [108] (p. 256), a step which we argue should include both native title claimants and holders in line with national water policy guidelines. Ideally, this would require an institutional response that departs from current ad hoc and manual processes and moves towards establishing and operating a clearer, though flexible, process to accommodate and protect native title rights to water [1]. As detailed above, however, this will necessarily be a complex, political and time consuming feat. We also agree with the suggestion from Mould and colleagues [105] for individuals within organizations to adjust their work styles where possible to place a stronger emphasis on informal dialog and relationships with landholders and water users, including Aboriginal peoples. This provides a means of practicing and influencing more holistic and collaborative water and river management, and potentially makes colonial vestiges of water management more visible and readily addressed. While these lessons apply particularly to the NSW Government, they could similarly improve planning practice in other Australian State and Territory Governments.

Beyond these changes to native title implementation and water planning practice and policy, we see the need for action on other fronts. As mentioned earlier, there is an increasing recognition that Aboriginal rights to water should include commercial rights. The Australian Law Reform Commission has handed down a report on reform to the *Native Title Act* in which it recommends changes that could see economic benefit accrue native title holders from the use of natural resources [66]. Any efforts to expand the currently narrow definitions of customary rights to water in either native title or water law will be welcome. While the Federal Government is yet to respond to this inquiry there is nonetheless considerable support for its recommendation amongst Aboriginal advocates [3,64,88,109]. Aboriginal lawyer Tony McAvoy sees such a reform as critical to Indigenous water rights struggles: 'real impact on the commercial market in water and therefore river management will only occur when Indigenous people are water owners themselves' [109] (p. 97). To that end, suggestions have been made for the establishment of an independent Indigenous Water Fund or Trust to allow Indigenous peoples to participate in the water market and allocate water to meet self-determined objectives [29]. Those funds could be used to purchase licences and would be managed to provide for necessary infrastructure and other costs associated with accessing water entitlements. Purchases and use costs could be funded on an ongoing basis from a small levy on water trades [109]. Under such an arrangement, Indigenous peoples might choose to direct water to the environment and could also pursue other water use strategies to underpin contemporary livelihoods [29].

In conclusion, it is these kinds of improvements that urgently need the attention of policy makers and practitioners. A fitting arena in which to debate and develop progressive change is the renegotiation of the National Water Initiative, raised as a realistic outcome for 2020 by the national body oversighting implementation of the current national policy [63]. There can be little doubt that unlike the first time national water policy recognized Indigenous rights and interests in water in 2004, the next time around, Indigenous people will at least be consulted. It is hoped that the recognition afforded Indigenous Australians in that most important exercise goes further than consultation and engages in good faith negotiation and thereby breaks with the pattern of misrecognition and non-recognition seen in recent decades.

Acknowledgments: We express our deepest gratitude to the Barkandji Traditional Owners interviewed in the course of this research, as well as NTSCORP and former NSW Government employees, especially Brad Moggridge (Kamilaroi Water Scientist) and Lillian Moseley, for their important contributions. We also wish to extend our appreciation to the two anonymous reviewers for their helpful suggestions and comments, all of which have improved the quality of this publication. Lana Hartwig was supported by an Australian Government Research Training Program (RTP) Scholarship and an Australian Rivers Institute Top-Up Scholarship. Sue Jackson was additionally supported by the Australian Research Council's Future Fellowships Program funding scheme (project number FT130101145).

Author Contributions: Lana Hartwig and Sue Jackson conceived and designed the research. Lana Hartwig conducted the fieldwork and analyzed the data, with Sue Jackson and Natalie Osborne substantially contributing supportive materials and analysis tools. All co-authors wrote the article.

Conflicts of Interest: The authors declare no conflict of interest.

References

1. Behrendt, J.; Thompson, P. The recognition and protection of Aboriginal interests in NSW rivers. *J. Indig. Policy* **2004**, *3*, 37–140.
2. Gibbs, L.M. Just add water: Colonisation, water governance and the Australian inland. *Environ. Plan. A* **2009**, *41*, 2964–2983. [CrossRef]
3. Jackson, S.; Langton, M. Trends in the recognition of Indigenous water needs in Australian water reform: The limitations of 'cultural' entitlements in achieving water equity. *J. Water Law* **2012**, *22*, 109–123.
4. Stephenson, M.A.; Ratnapala, S. (Eds.) *Mabo, a Judicial Revolution: The Aboriginal Land Rights Decision and Its Impact on Australian Law*; University of Queensland Press: Brisbane, Australia, 1993.
5. Brennan, S.; Davis, M.; Edgeworth, B.; Terrill, L. (Eds.) *Native Title from Mabo to Akiba: A Vehicle for Change and Empowerment?* Federation Press: Sydney, Australia, 2015.
6. Commonwealth of Australia and the Governments of New South Wales, Victoria, Queensland, South Australia, the Australian Capital Territory and the Northern Territory. *Intergovernmental Agreement on a National Water Initiative*; National Water Commission: Canberra, Australia, 2004.
7. Perreault, T. Custom and contradiction: Rural water governance and the politics of *usos y costumbres* in Bolivia's irrigators' movement. *Ann. Assoc. Am. Geogr.* **2008**, *98*, 834–854. [CrossRef]
8. Robison, J.; Cosens, B.; Jackson, S.; Leonard, K.; McCool, D. Indigenous water justice. *Lewis Clark Law Rev.* forthcoming.
9. Altman, J.; Markham, F. Burgeoning Indigenous land ownership: Diverse values and strategic potentialities. In *Native Title from Mabo to Akiba: A Vehicle for Change and Empowerment?* Brennan, S., Davis, M., Edgeworth, B., Terrill, L., Eds.; Federation Press: Sydney, Australia, 2015; pp. 126–142.
10. Jackson, S.; Tan, P.-L.; Mooney, C.; Hoverman, S.; White, I. Principles and guidelines for good practice in Indigenous engagement in water planning. *J. Hydrol.* **2012**, *474*, 57–65. [CrossRef]

11. Tan, P.-L.; Jackson, S. Impossible dreaming—Does Australia's water law and policy fulfil Indigenous aspirations? *Environ. Plan. Law J.* **2013**, *30*, 132–149.

12. O'Donnell, M. The National Water Initiative, native title rights to water and the emergent recognition of Indigenous specific commercial rights to water in Northern Australia. *Australas. J. Nat. Res. Law Policy* **2013**, *16*, 83–100.

13. Weir, J. *Murray River Country: An Ecological Dialogue with Traditional Owners*; Aboriginal Studies Press: Canberra, Australia, 2009.

14. Ivison, D. Justification, not recognition. *Indig. Law Bull.* **2016**, *8*, 12–18.

15. Hay, C.; Lister, M. Introduction: Theories of the state. In *The State: Theories and Issues*; Hay, C., Lister, M., Marsh, D., Eds.; Palgrave Macmillan: Hampshire, UK, 2006; pp. 1–20.

16. Scott, J.C. *Seeing Like a State: How Certain Schemes to Improve the Human Condition Have Failed*; Yale University Press: New Haven, CT, USA, 1998.

17. Porter, L. Rights or containment? The politics of Aboriginal cultural heritage in Victoria. *Aust. Geogr.* **2006**, *37*, 355–374. [CrossRef]

18. Povinelli, E.A. *The Cunning of Recognition: Indigenous Alterities and the Making of Australian Multiculturalism*; Duke University Press: Durham, NC, USA, 2002.

19. Povinelli, E.A. Radical worlds: The anthropology of incommensurability and inconceivability. *Annu. Rev. Anthrop.* **2001**, *30*, 319–334. [CrossRef]

20. Bakker, K. Water: Political, biopolitical, material. *Soc. Stud. Sci.* **2012**, *42*, 616–623. [CrossRef]

21. Fraser, N. Rethinking recognition. *New Left Rev.* **2000**, *3*, 107–120.

22. Kowal, E.; Paradies, Y. Indigeneity and the refusal of whiteness. *Postcolon. Stud.* **2017**, *20*, 101–117. [CrossRef]

23. Hunt, S. Ontologies of Indigeneity: The politics of embodying a concept. *Cult. Geogr.* **2014**, *21*, 27–32. [CrossRef]

24. Simpson, A. The ruse of consent and the anatomy of 'refusal': Cases from indigenous North America and Australia. *Postcolon. Stud.* **2017**, *20*, 18–33. [CrossRef]

25. Wolfe, P. Land, labor and difference: Elementary structures of race. *Am. Hist. Rev.* **2001**, *106*, 866–905. [CrossRef]

26. Wolfe, P. Settler colonialism and the elimination of the native. *J. Genocide Res.* **2006**, *8*, 387–409. [CrossRef]

27. Balaton-Chrimes, S.; Stead, V. Recognition, power and coloniality. *Postcolon. Stud.* **2017**, *20*, 1–17. [CrossRef]

28. Ivison, D. Pluralising political legitimacy. *Postcolon. Stud.* **2017**, *20*, 118–130. [CrossRef]

29. Macpherson, E. Beyond recognition: Lessons from Chile for allocating Indigenous water rights in Australia. *Univ. N. S. W. Law J.* **2017**, *40*, 1130–1169.

30. Tully, J. Recognition and dialogue: The emergence of a new field. *Crit. Rev. Int. Soc. Political Philos.* **2004**, *7*, 84–106. [CrossRef]

31. Boelens, R. Local rights and legal recognition: The struggle for Indigenous water rights and the cultural politics of participation. In *Water and Indigenous Peoples*; Boelens, R., Chiba, M., Nakashima, D., Eds.; United Nations Educational, Scientific and Cultural Organization (UNESCO): Paris, France, 2006; pp. 46–61.

32. Boelens, R.; Bustamante, R.; de Vos, H. Legal pluralism and the politics of inclusion: Recognition and contestation of local water rights in the Andes. In *Community-Based Water Law and Water Resource Management Reform in Developing Countries*; van Koppen, B., Giordano, M., Butterworth, J., Eds.; Centre for Agriculture and Bioscience International (CABI): Cambridge, MA, USA, 2007; pp. 96–113.

33. Boelens, R. The politics of disciplining water rights. *Dev. Chang.* **2009**, *40*, 307–331. [CrossRef]

34. Coulthard, G.S. Subjects of empire: Indigenous peoples and the 'politics of recognition' in Canada. *Contemp. Political Theory* **2007**, *6*, 437–460. [CrossRef]

35. Strelein, L.; Tran, T. Building Indigenous governance from native title: Moving away from 'fitting in' to creating a decolonized space. *Rev. Const. Stud.* **2013**, *18*, 19–48.

36. Watson, I. The future is our past: We once were sovereign and we still are. *Indig. Law Bull.* **2012**, *8*, 12–15.

37. Referendum Council. Uluru Statement from the Heart: Statement of the First Nations National Constitutional Convention. Available online: https://www.referendumcouncil.org.au/sites/default/files/2017-5/Uluru_Statement_From_The_Heart_0.PDF (accessed on 31 January 2018).

38. Martin, S. *Aboriginal Cultural Heritage for the Menindee Lakes Area: Part 1 Aboriginal Ties to the Land*; A Report to the Menindee Lakes Ecologically Sustainable Development Project; Steering Committee: Broken Hill, Australia, 2001.

39. Hardy, B. *Lament for the Barkindji: The Vanished Tribes of the Darling River Region*; Rigby Limited: Adelaide, Australia, 1976.

40. Heritage Office and Department of Urban Affairs & Planning. *Regional Histories: Regional Histories of New South Wales [Online]*; Heritage Office: Sydney, Australia, 1996; pp. 190–211. Available online: http://www.environment.nsw.gov.au/resources/heritagebranch/heritage/RegionalHistoriesPt7WestPlainsLordHoweIsAppendices.pdf (accessed on 6 November 2017).

41. Jackson, S. Enduring and persistent injustices in water access in Australia. In *Natural Resources and Environmental Justice: The Australian Experience*; Lukasiewicz, A., Dovers, S., Robin, J., McKay, J., Schilizzi, S., Graham, S., Eds.; Commonwealth Scientific and Industrial Research Organisation (CSIRO) Publishing: Melbourne, Australia, 2017; pp. 121–132.

42. Burdon, P.; Drew, G.; Stubbs, M.; Webster, A.; Barber, M. Decolonising Indigenous water 'rights' in Australia: Flow, difference, and the limits of law. *Settl. Colon. Stud.* **2015**, *5*, 334–349. [CrossRef]

43. Forsyth, H.; Gavranovic, A. The logic of survival: Towards an Indigenous-centred history of capitalism in Wilcannia. *Settl. Colon. Stud.* **2017**, *1*, 1–25. [CrossRef]

44. McCausland, R.; Vivian, A. Why do some Aboriginal communities have lower crime rates than others? A pilot study. *Aust. N. Z. J. Criminol.* **2010**, *43*, 301–332. [CrossRef]

45. Gibson, L. 'We are the river': Place, wellbeing and Aboriginal identity. In *Wellbeing and Place*; Atkinson, S., Fuller, S., Painter, J., Eds.; Ashgate: Surrey, UK, 2012; pp. 201–215.

46. Barkandji Traditional Owners #8 v Attorney-General of New South Wales [2015] FCA 604. Available online: http://www.austlii.edu.au/cgi-bin/viewdoc/au/cases/cth/FCA/2015/604.html (accessed on 24 October 2017).

47. Breen, J.; Coote, G. Largest native title claim in NSW acknowledges Barkandji people in state's far west. *ABC News*, 16 June 2015. Available online: http://www.abc.net.au/news/2015-06-16/nsw-largest-native-title-claim-determination/6549180 (accessed on 8 November 2017).

48. Pardoe, C. The Menindee Lakes: A regional archaeology. *Aust. Archaeol.* **2003**, *57*, 42–53. [CrossRef]

49. Murray Darling Basin Authority. *Lower Darling Reach Report, Constraints Management Strategy*; Murray Darling Basin Authority: Canberra, Australia, 2015. Available online: https://www.mdba.gov.au/sites/default/files/pubs/Lower-Darling-reach-report-aug-2015.pdf (accessed on 24 October 2017).

50. ABC. Pumped: Who's benefitting from the billions spent on the Murray-Darling? *Four Corners*, 24 July 2017. Available online: http://www.abc.net.au/4corners/pumped/8727826 (accessed on 24 October 2017).

51. Brewster, K. Murray-Darling Basin: Goondiwindi cotton farm raided over alleged misuse of Commonwealth funding. *Lateline via ABC News*. 26 October 2017. Available online: http://www.abc.net.au/news/2017-10-24/murray-darling-basin-alleged-fraud-goondiwindi-cotton-farm/9080280 (accessed on 9 November 2017).

52. Matthews, K. *Independent Investigation into NSW Water Management and Compliance. Interim Report*; NSW Department of Industry: Sydney, Australia, 2017. Available online: https://www.industry.nsw.gov.au/__data/assets/pdf_file/0016/120193/Matthews-interim-report-nsw-water.pdf (accessed on 24 October 2017).

53. Atkinson, R.; Flint, J. Accessing hidden and hard-to-reach populations: Snowball research strategies. *Soc. Res. Update* **2001**, *33*, 1–4.

54. Bates, W. When they take the water from a Barkandji person, they take our blood. *The Guardian*, 26 July 2017. Available online: https://www.theguardian.com/commentisfree/2017/jul/26/when-they-take-the-water-from-a-barkandji-person-they-take-our-blood (accessed on 8 November 2017).

55. Wainwright, S.; Gooch, D. Far west NSW campaigners staying at Canberra tent embassy to demand better water management. *ABC Broken Hill*. 2 May 2016. Available online: http://www.abc.net.au/news/2016-05-02/far-west-nsw-campaigners-canberra-over-water/7376288 (accessed on 26 October 2017).

56. Wainwright, S. Indigenous protestors march to show their anger over the drying up of the Darling River. *ABC Broken Hill*. 20 June 2016. Available online: http://www.abc.net.au/news/2016-06-20/indigenous-protesters-protest-over-state-of-darling-river/7526718 (accessed on 26 October 2017).

57. Weir, J. Water planning and dispossession. In *Basin Futures: Water Reform in the Murray-Darling Basin*; Connell, D., Grafton, R.Q., Eds.; Australian National University Electronic (ANUE) Press: Canberra, Australia, 2011; pp. 179–191.

58. Connell, D.; Robins, L.; Dovers, S. Delivering the National Water Initiative. In *Managing Water for Australia: The Social and Institutional Challenges*; Hussey, K., Dovers, S., Eds.; CSIRO Publishing: Melbourne, Australia, 2007; pp. 127–140.

59. Jackson, S.; Morrison, J. Indigenous perspectives in water management, reforms and implementation. In *Managing Water for Australia: The Social and Institutional Challenges*; Hussey, K., Dovers, S., Eds.; CSIRO Publishing: Melbourne, Australia, 2007; pp. 23–41.

60. National Water Commission. *Australian Water Reform 2009: Second Biennial Assessment of Progress in Implementation of the National Water Initiative*; National Water Commission: Canberra, Australia, 2009.

61. National Water Commission. *The National Water Initiative—Securing Australia's Water Future: 2011 Assessment*; National Water Commission: Canberra, Australia, 2011.

62. National Water Commission. *A Review of Indigenous Involvement in Water Planning, 2013*; National Water Commission: Canberra, Australia, 2014.

63. Productivity Commission. *National Water Reform Inquiry, Draft Report*; Productivity Commission: Canberra, Australia, 2017.

64. Marshall, V. *Overturning Aqua Nullius: Securing Aboriginal Water Rights*; Aboriginal Studies Press: Canberra, Australia, 2017.

65. Jackson, S. Planning in the Native Title era. In *Planning in Indigenous Australia: From Imperial Foundations to Postcolonial Futures*; Jackson, S., Porter, L., Johnson, L.C., Eds.; Routledge: New York, NY, USA, 2017; pp. 175–194.

66. Australian Law Reform Commission. *Connection to Country: Review of the Native Title Act 1993 (Cth)*; ALRC Report 126; Australian Law Reform Commission: Sydney, Australia, 2015.

67. Strelein, L. The right to resources and the right to trade. In *Native Title from Mabo to Akiba: A Vehicle for Change and Empowerment?* Brennan, S., Davis, M., Edgeworth, B., Terrill, L., Eds.; Federation Press: Sydney, Australia, 2015; pp. 44–59.

68. National Native Title Tribunal. Statistics. Available online: http://www.nntt.gov.au/Pages/Statistics.aspx (accessed on 22 January 2018).

69. Hunt, J.; Ellsmore, S. Navigating a path through delays and destruction: Aboriginal cultural heritage protection in New South Wales using native title and land rights. In *The Right to Protect Sites: Indigenous Heritage in the Era of Native Title*; McGrath, P.F., Ed.; Australian Institute of Aboriginal and Torres Strait Islander Studies (AIATSIS) Research Publications: Canberra, Australia, 2016; pp. 77–110.

70. Altman, J. Indigenous interests and water property rights. *Dialogue* **2004**, *23*, 29–34.

71. Morgan, M.; Strelein, L.; Weir, J. Indigenous water rights within the Murray Darling Basin. *Indig. Law Bull.* **2004**, *5*, 17–20.

72. Duff, N. *Fluid Mechanics: The Practical Use of Native Title for Freshwater Outcomes*; AIATSIS Research Publications: Canberra, Australia, 2017.

73. O'Donnell, M. *Indigenous Rights in Water in Northern Australia, NAILSMA TRaCK Project 6.2*; Charles Darwin University: Darwin, Australia, 2011.

74. Bartlett, R.H. *Native Title in Australia*, 3rd ed.; LexisNexis Butterworths: Chatswood, Australia, 2015.

75. O'Bryan, K. More Aqua Nullius: The *Traditional Owner Settlement Act 2010* (Vic) and the neglect of Indigenous rights to manage inland water resources. *Melb. Univ. Law Rev.* **2016**, *40*, 547–593.

76. Bell, S.; Park, A. The problematic metagovernance of networks: Water reform in New South Wales. *J. Public Policy* **2006**, *26*, 63–83. [CrossRef]

77. NSW Office of Water. *Our Water Our Country: An Information Manual for Aboriginal People and Communities about the Water Reform Process*; NSW Department of Primary Industries: Sydney, Australia, 2012. Available online: http://www.webcitation.org/6v9b4wCkN (accessed on 22 November 2017).

78. Tan, P.-L. *Collaborative Water Planning: Legal and Policy Analysis. Report to the Tropical Rivers and Coastal Knowledge (TRaCK) Program*; Land & Water Australia: Canberra, Australia, 2008.

79. Hamstead, M.; Baldwin, C.; O'Keefe, V. *Water Allocation Planning in Australia—Current Practices and Lessons Learned. Waterlines Occasional Paper No. 6*; National Water Commission: Canberra, Australia, 2008.

80. NSW DPI—Water. Stakeholder Advisory Panel: Terms of Reference. May 2017. Available online: http://www.water.nsw.gov.au/__data/assets/pdf_file/0006/735198/Stakeholder-advisory-panel-terms-of-reference.pdf (accessed on 22 November 2017).

81. NSW DPI—Water. NSW Groundwater Water Resource Plans: Stakeholder Advisory Panel Terms of Reference. August 2017. Available online: http://www.water.nsw.gov.au/__data/assets/pdf_file/0011/736580/Groundwater-SAP-terms-of-reference.pdf (accessed on 22 November 2017).

82. NSW Department of Industry. The Basin Plan for the Murray-Darling. Available online: https://www.water.nsw.gov.au/water-management/law-and-policy/national-reforms/murray-darling-basin-plan (accessed on 31 January 2018).

83. NSW Department of Industry. Submission on the Productivity Commission's Draft Report on National Water Reform. October 2017. Available online: https://www.pc.gov.au/__data/assets/pdf_file/0009/222777/subdr116-water-reform.pdf (accessed on 22 November 2017).

84. Water Sharing Plan for the Barwon-Darling Unregulated and Alluvial Water Sources. 2012. Available online: https://legislation.nsw.gov.au/#/view/subordleg/2012/488 (accessed on 24 October 2017).

85. NSW DPI. Replacement, Merged and New Water Sharing Plans Commenced. 1 July 2016. Available online: http://www.water.nsw.gov.au/water-management/water-sharing/replacement-merged-new (accessed on 22 November 2017).

86. NSW DPI. Water Sharing Plans. Available online: http://www.water.nsw.gov.au/water-management/water-sharing (accessed on 22 November 2017).

87. Murrumbidgee Groundwater Preservation Association Inc. v Minister for Natural Resources [2005] NSWCA 10. Available online: https://www.austlii.edu.au/cgi-bin/viewdoc/au/cases/nsw/NSWCA/2005/10.html (accessed on 23 January 2018).

88. Taylor, K.S.; Moggridge, B.J.; Poelina, A. Australian Indigenous Water Policy and the impacts of the ever-changing political cycle. *Australas. J. Water Res.* **2017**, *20*, 132–147. [CrossRef]

89. Tan, P.-L.; Bowmer, K.H.; Mackenzie, J. Deliberative tools for meeting the challenges of water planning in Australia. *J. Hydrol.* **2012**, *474*, 2–10. [CrossRef]

90. Water Sharing Plan for the Clarence River Unregulated and Alluvial Water Sources. 2016. Available online: https://legislation.nsw.gov.au/#/view/subordleg/2016/381 (accessed on 24 October 2017).

91. Water Sharing Plan for the Greater Metropolitan Region Unregulated River Water Sources. 2011. Available online: https://www.legislation.nsw.gov.au/#/view/regulation/2011/112 (accessed on 24 October 2017).

92. General Purpose Standing Committee No 5 NSW Parliament Legislative Council. Inquiry into the augmentation of Water Supply for Rural and Regional New South Wales: Proceedings at Broken Hill. 26 October 2016. Available online: https://www.parliament.nsw.gov.au/committees/DBAssets/InquiryEventTranscript/Transcript/9868/Transcript%20-%2026%20October%202016%20-%20Corrected.pdf (accessed on 22 November 2017).

93. Water Sharing Plan for the NSW Murray Darling Basin Fractured Rock Groundwater Sources. 2011. Available online: https://legislation.nsw.gov.au/#/view/regulation/2011/615/id6 (accessed on 24 October 2017).

94. Water Sharing Plan for the NSW Murray Darling Basin Porous Rock Groundwater Sources. 2011. Available online: https://legislation.nsw.gov.au/#/view/regulation/2011/616/id7 (accessed on 24 October 2017).

95. Water Sharing Plan for the Lower Murray-Darling Unregulated and Alluvial Water Sources. 2011. Available online: https://www.legislation.nsw.gov.au/#/view/subordleg/2012/22 (accessed on 24 October 2017).

96. Water Sharing Plan for the New South Wales Murray and Lower Darling Regulated Rivers Water Sources. 2016. Available online: https://legislation.nsw.gov.au/#/view/subordleg/2016/366 (accessed on 24 October 2017).

97. NSW DPI. NSW Murray and Lower Darling Regulated Rivers. Available online: http://www.water.nsw.gov.au/water-management/water-sharing/plans_commenced/water-source/nsw-mld-reg-rivers (accessed on 22 November 2017).

98. NSW DPI. Summary of Consultation Undertaken for Inland Water Sharing Plans That Were Replaced on 1 July 2016. Available online: http://www.water.nsw.gov.au/__data/assets/pdf_file/0011/660539/Summary-of-inland-consultation.pdf (accessed on 22 November 2017).

99. Karpany v Dietman [2013] HCA 47. Available online: https://www.austlii.edu.au/cgi-bin/viewdoc/au/cases/cth/HCA/2013/47.html (accessed on 23 January 2018).

100. Adcock, B. Aboriginal fishing: When culture becomes criminal. *Background Briefing ABC*. 21 February 2016. Available online: http://www.abc.net.au/radionational/programs/backgroundbriefing/aboriginal-fishing:-when-culture-becomes-criminal/7180772 (accessed on 23 January 2018).

101. NTSCORP Limited. Supplementary Questions for NTSCORP; General Purpose Standing Committee No 5 NSW Parliament Legislative Council, Inquiry into the Augmentation of Water Supply for Rural and Regional New South Wales: 26 October 2016. Available online: https://www.parliament.nsw.gov.au/committees/DBAssets/InquiryOther/Transcript/10582/Answers%20to%20supplementary%20questions%20-%20NTSCORP.pdf (accessed on 22 November 2017).

102. NTSCORP Limited. Inquiry into Water Augmentation. Submission No. 84. General Purpose Standing Committee No 5 NSW Parliament Legislative Council, Inquiry into the Augmentation of Water Supply for Rural and Regional New South Wales: 15 August 2016. Available online: https://www.parliament.nsw.gov.au/committees/DBAssets/InquirySubmission/Body/56268/084%20NTSCORP.pdf (accessed on 22 November 2017).

103. Jackson, S. Compartmentalising culture: The articulation and consideration of Indigenous values in water resource management. *Aust. Geogr.* **2006**, *37*, 19–31. [CrossRef]

104. Murray Lower Darling Rivers Indigenous Nations. Echuca Declaration. Available online: http://www.savanna.org.au/nailsma/publications/downloads/MLDRIN-NBAN-ECHUCA-DECLARATION-2009.pdf (accessed on 24 November 2017).

105. Mould, S.A.; Fryirs, K.; Howitt, R. Practicing Sociogeomorphology: Relationships and dialog in river research and management. *Soc. Nat. Res.* **2017**, *31*, 106–120. [CrossRef]

106. Suchet, S. 'Totally Wild'? Colonising discourses, indigenous knowledges and managing wildlife. *Aust. Geogr.* **2002**, *33*, 141–157. [CrossRef]

107. Western Bundjalung People v Attorney General of New South Wales [2017] FCA 992. Available online: https://www.austlii.edu.au/cgi-bin/viewdoc/au/cases/cth/FCA/2017/992.html (accessed on 23 January 2018).

108. Holley, C. Future water: Improving planning, markets, enforcement and learning. In *New Directions for Law in Australia: Essays in Contemporary Law Reform*; Levy, R., O'Brian, M., Rice, S., Ridge, P., Thornton, M., Eds.; ANU Press: Canberra, Australia, 2017; pp. 253–262.

109. McAvoy, T. Water—Fluid perceptions. *Transform. Cult. J.* **2006**, *1*, 97–103. [CrossRef]

resources

MDPI

Article

An Equity Autopsy: Exploring the Role of Water Rights in Water Allocations and Impacts for the Central Valley Project during the 2012–2016 California Drought

Zachary P. Sugg [1,2]

[1]   Martin Daniel Gould Center for Conflict Resolution, Stanford Law School, Stanford University, Stanford, CA 94305, USA; zachary.sugg@stanford.edu
[2]   Water in the West Program, Woods Institute for the Environment , Stanford University, Stanford, CA 94305, USA

Received: 27 November 2017; Accepted: 6 February 2018; Published: 13 February 2018

Abstract: Entrenched Western water rights regimes may appear to function relatively well in wet years, but extreme drought events can expose the kinds of harsh ecological and socio-economic outcomes that the hard edges of prior appropriation inherently generate. During the 2012–2016 California drought some irrigators received little or no water at all in consecutive years while others received comparatively large allocations. This paper focuses on the role that California's water rights priority system and its administration via Central Valley Project contracts have played in generating disproportionate water allocations and impacts during the drought. The analysis is structured around two key questions: (a) in what ways does strict adherence to a priority system of water allocations produce inequitable socio-ecological outcomes during severe drought? (b) how might the system be changed to foster outcomes that are more equitable and fair, and with less costly and less serious conflicts in a non-stationary climate future marked by extreme events? Using an equity perspective, I draw from the doctrine of equitable apportionment to imagine a water rights regime that is better able to create a fairer distribution of drought impacts while meaningfully elevating the importance of future generations and increasing adaptive capacity.

Keywords: water law; California; drought; equity; Central Valley Project; water rights; prior appropriation

---

1. Introduction

From 2012–2016 California experienced the most severe drought—including the driest single year, 2014—in the last 1200 years [1]. Statewide costs to California's agricultural sector in 2015 were estimated at $2.7 billion, along with some 21,000 workers impacted by either direct or indirect job losses [2]. Agriculture largely rode out the drought by continuing to deplete the already vastly overdrafted groundwater, but at a cost of $590 million in that year alone. Statewide figures mask the uneven socioeconomic impacts of the drought, which fell disproportionately on agricultural areas south of the Sacramento/San Joaquin Bay Delta in the San Joaquin Valley [3–5].

An important factor affecting these impacts was curtailment of surface water supplies from federal managers, with many growers receiving 0–5% of their contracted maximum allocation for three consecutive years [6]. Much of California's surface water is managed by the federal Central Valley Project (CVP), an expansive system of 20 reservoirs, 11 powerplants, and over 500 miles of canals that facilitates interbasin transfers from the Sacramento River Basin south into certain parts of the relatively drier San Joaquin Basin [7]. USBR administers CVP water supply contracts totaling to some 9.5 million acre-feet (AF), though average actual deliveries are about 7 million AF [8]. About 5 million AF is used to irrigate about one-third of all the agricultural land in the state, with the rest divided up between municipal and industrial (M&I) uses, in-stream flows, and wildlife refuges and wetlands [8]

CVP allocations are shaped by the underlying state water rights system which, while a hybrid system of riparian and prior appropriation rights, has a defined hierarchy based largely on chronology of right establishment [9]. CVP contracts reflect this hierarchy and dictate to some extent how the negative impacts of shortages are distributed among different groups of contract holders [9,10].

While water rights are only one factor affecting "acquisitions and allocations" within the universe of CVP contractors, they take on especially significant importance during times of relative scarcity [10]. This particular drought was severe enough that observers began broaching the sacrosanct third rail of Western water: the prior appropriation system [11–18].

Although this drought was unprecedented in the last millennium, tree ring records show that there have been longer and more severe natural droughts within the climate system [1]. Given robust modeling estimates of increasingly higher mean and extreme temperatures [19], reduced snowpack [20], and increased risk of droughts that are both hot and dry [21], it would be a mistake to fail to attempt to learn from it.

Western states like California must adapt to new climate realities in ways that do not exacerbate already existing inequities or create new ones. Recent studies of equity and drought in California have focused on the uneven access to domestic water supplies in marginalized communities [22,23]. In this analysis I seek to expand such analyses by turning the equity lens to the state's water rights system and how it dictates uneven outcomes for communities of water users who receive water from the CVP. I ask two questions: (1) in what ways does strict adherence to a priority system of water allocations produce inequitable socio-ecological outcomes during severe drought? (2) how might the system be changed to foster outcomes that are more equitable and fair, and with less costly and less serious conflicts in a non-stationary climate future marked by extreme events?

I argue that evidence from the drought supports critiques of the fairness of the priority system of water allocation and that the equitable apportionment approach would be an alternative system more conducive to meaningfully incorporating equity dimensions of drought response. I focus in particular on how severe curtailments to CVP allocations exacerbate the groundwater overdraft problem in the Central Valley. The paper is organized in the following sections: (1) introduction; (2) short review of literature on water resource equity; (3) historical and legal background for CVP contracts; (4) analysis of the 2012–2016 drought and impacts for the CVP contractors; (5) discussion (6) conclusion.

2. Equity and Water Resources in the West and Beyond

In the context of the Western U.S., the "equity perspective" arose amid the broader shift from large water storage projects to water transfers in the 1980s after federal support for dams and diversions tapered off during and after the Carter administration [24]. Some scholars recognized in the 1980s that subjecting control and access to surface water supplies to a market economy had strong potential to exacerbate already unfair water access among Indian tribes and traditional Hispano communities throughout the Southwest for whom treating water as a tradable commodity has been considered morally unacceptable [25].

How can the equity perspective be characterized in the context of water resources? Ingram et al. [26] (p. 6) contend that equity has no single universal objective definition, but rather is more properly understood as

> " . . . *complex and contingent on circumstances, varied and nuanced, and cannot be fully understood until put back into the life cycle of living things. Consequently, there is no simple principle or set of principles, like those guiding efficiency, which can be set out as rules and universally applied in all places and circumstances. Instead, equity is a complex and protean idea."*

Many western water observers know that in the West, equity is meted out very differently within states compared to among states. Within states, the seniority system dominates, while interstate agreements usually rest on the doctrine of equitable apportionment [27,28]. Similarly, water markets may be seen by officials as a fair allocation mechanism to use within states, but unacceptable at the interstate level [29].

Because equity is a somewhat nebulous concept, scholars have investigated empirically how equity has been operationalized and acted out by communities of water users in a range of contexts, including poor water access in indigenous and Hispano communities in the U.S. Southwest [25]; a wide variety of case studies across Mexico, Spain, the American Southwest, and Pacific Northwest [30]; and the Global South [31–35]. Recent work has also explored the "climate gap" between disadvantaged and more affluent communities in relation to the degree of vulnerability to climate change and resources for mitigation and adaptation in communities near the Arizona and New Mexico border [36]. Much of this recent work weaves equity concerns into broader discourses about water justice and climate justice.

During the 1990s the Australian government funded a series of seven social scientific studies spanning a decade to determine what exactly people think constitutes equity and fairness in water allocation. The researchers found that " . . . people have universal fairness criteria for judging the overall fairness of water allocation systems at a general level, and these are useful for systematic derivation of accountable solutions at a local or situational level," though may shift towards greater emphasis on situation-specific fairness criteria depending on the perceived urgency of a problem [37] (p. 67).

More empirical work is needed to elicit social understandings of equity in different water user communities in the Western U.S.; a long-term study like the Australian one just mentioned to systematically determine the understandings of water allocation equity across the various stakeholder and user groups connected to the CVP would be particularly valuable. However, in the absence of such a study, an operational definition of equity is needed for the sake of argument and clarity. I borrow Dunning's [38] (p. 77) definition of equity as "an attempt to fairly share limited water resources, often by taking into considerations many factors".

More definitional specificity is required to meaningfully evaluate a given case study using the lens of equity, but this is challenging for at least two reasons: (1) equity is always situation-specific to some extent, and (2) there is not just one kind of equity but rather multiple forms or types (socioeconomic, procedural, intergenerational, etc.). Rather than argue for a single definition, scholars have specified the meaning of equity in terms of sets of principles [39,40]. Constitutive principles can be understood as necessary and sufficient conditions, a qualitative balancing test for evaluating equity in water policy [40] (p. 186).

To further clarify the operational meaning of equity for this analysis I draw from the five equity principles articulated by Ingram et al. [40] in their analysis of the allocation of the Colorado River.

1. *Reciprocity* means "distributive advantages and costs should be shared by all members of the relevant community" [40] (p. 186). It is a balancing principle that recognizes the fairness of prior appropriation in its original 19th century setting, while also recognizing that in certain conditions it can result in intolerable effects, such as waste and inefficiency resulting from the "use it or lose it" provision, or harm to the rights of third parties.

2. *Value pluralism* means "users' rights to employ water to pursue whatever values they consider legitimate should be respected, provided use does not degrade the resource or harm others" [40] (p. 187). Ingram et al. recognize that the conditions of no degradation and no harm could restrict certain activities and uses of water and thus must be balanced by the principle of reciprocity.

3. Principle 3 is *ensuring the accommodation of multiple value claims in resource allocation and decision processes*. This principle entails widening the diversity of communities involved in decisions and rejects sacrificing participation in the pursuit of technically efficient decisions, even if it makes deliberation and decision-making messier and more complicated. For Ingram et al., such inconveniences ought to be tolerated because the alternatives are even less likely to satisfy the public interest in water resource decisions.

4. Principle 4 is to *obey promises agreed to in good faith*. Past negotiated agreements about the apportionment of water resources should be respected to the extent possible. Two special

problems with this principle are that (a) promises can conflict with each other, and (b) the circumstances under which promises were made can change over time to such an extent that the original agreements become highly problematic. Since there is no single "unambiguous rule of equity" for resolving such conflicts, flexibility, adaptation, and the acceptance of unavoidable ambiguity in decision making are especially important [40] (pp. 188–189). Any renegotiation of contracts in light of changed circumstances must be qualified by the other four equity principles.

5. *Intergenerational equity* is the principle that "the present use of water resources should take account of future generations" [40] (p. 189). Importantly, because water is fundamentally a social good, intergenerational equity is an inherently value-laden, ethical idea and thus cannot be satisfied by relying only on economic logics which rationalize the risks that short-term depletion and degradation may pose to future generations [39–42]. Intergenerational inequity is inextricably tied to sustainability [43–46] and is especially elevated in importance by climate change, as the most pronounced effects will be visited upon generations not yet born.

3. Background: How the Priority System is Embedded in Central Valley Project Water Supply Contracts

Understanding water impacts of drought in California requires also understanding the contracts water users have with the federal government through the CVP. In the 1920s, a basic issue for the state was that none of the then-recent actions (e.g., regulating post-1914 water rights, adjudications, creating irrigation districts) had generated any "new" water [47]. Political support had grown for a large storage project to significantly expand agriculture on the fertile but unirrigated soils of the Central Valley. Originally a state project approved by California voters, the state could not sell the bonds it needed to fund construction of the CVP during the Depression and it was taken over by the U.S. Bureau of Reclamation (USBR).

The project involved a complex system of dams and canals to store and divert water from the relatively water-rich Sacramento River Valley south to the drier San Joaquin Valley, and from the San Joaquin headwaters mainly to growers in the southern part of the Valley. A huge volume of water rights was needed to make the two massive transfers possible but both rivers were already grossly over-appropriated and the water rights to those rivers were largely very senior, monopolized, and unregulated. A major problem, therefore, was how to get adequate and secure rights to make the project legally and operationally feasible.

Two main things occurred to make it work. First, the state made major filings for new water rights for itself in 1927 and assigned them to USBR. However, since a priority date of 1927 was far junior to the existing users who collectively already had rights to more than twice the natural flows of both rivers, they also had to do something to make their rights more secure. Rather than push the state for adjudications of the San Joaquin and Sacramento Rivers, USBR opted to negotiate with the senior rights holders in both basins.

There are three main groups of CVP irrigation contractors–San Joaquin River Exchange Contractors, Sacramento River Settlement Contractors, and water service contracts, with each falling somewhere in the larger hierarchy of state water rights and CVP contracts in this general order:

(1) Riparian rights (includes Settlement and Exchange Contractors)
(2) Pre-1914 appropriation rights (rights acquired before state regulation)
(3) Post-1914 appropriation rights (rights acquired after state regulation, including USBR's rights for the CVP storage)
(4) CVP water service contracts

The geographic locations of the different CVP divisions and user groups are shown in Figure 1.

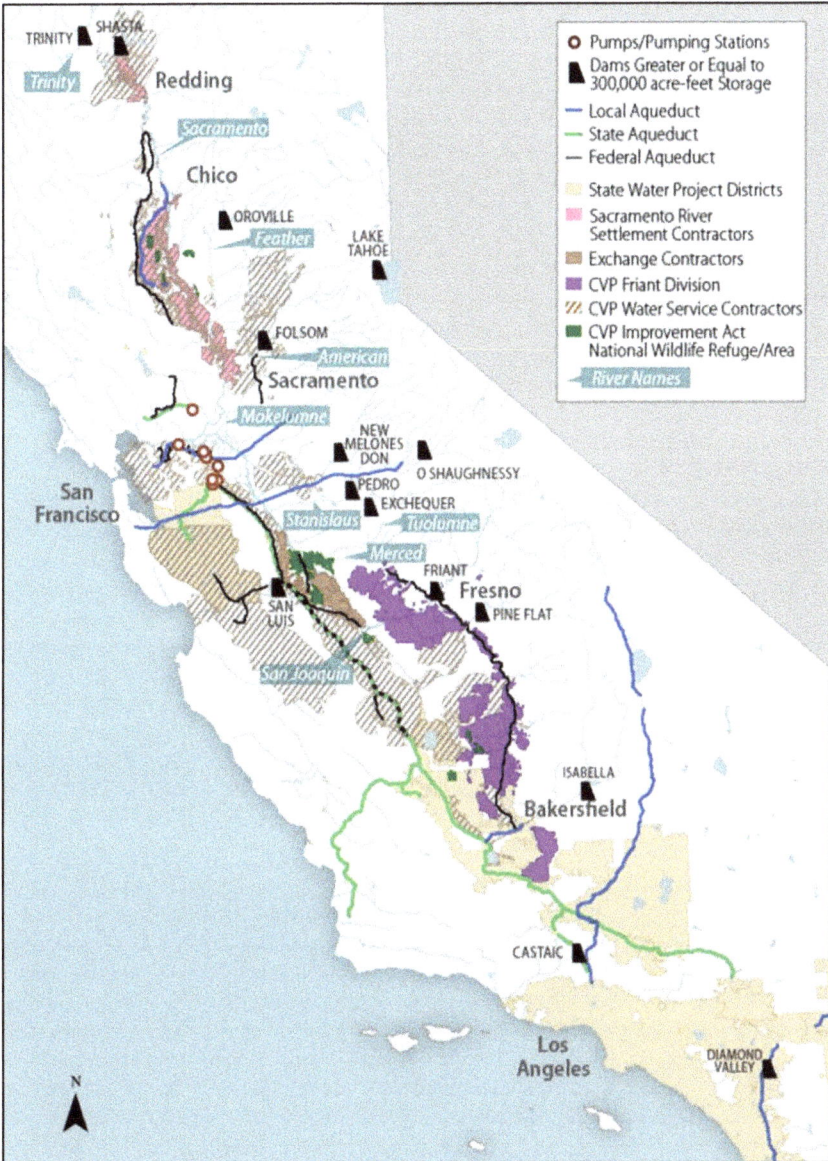

Figure 1. Geographical distribution of CVP and State Water Project infrastructure and water supply contractor divisions. Map modified from Cody et al. [10] (p. 36).

3.1. San Joaquin River Purchase and Exchange Contracts

The most senior water users within the CVP universe are the so-called Exchange Contractors who are the corporate descendants of the old Miller & Lux cattle company. The CVP plan entailed damming and diverting practically the entire flow of the San Joaquin River north and south along the Madera and Friant-Kern Canals (Figure 1), but this could not be done without infringing on the downstream riparian rights of Miller & Lux. Rather than pursue a water rights adjudication, USBR opted to strike an agreement to provide the senior San Joaquin diverters with a substitute supply from the Sacramento River in exchange for permission to store and divert the water they would otherwise use at Friant Dam.

In a Purchase Agreement signed in 1939, USBR acquired outright 623,000 AF from Miller & Lux, a mixture of riparian flood flows and pre-1914 appropriation rights to waters that flow through grasslands. This water would go to farmers along the Madera and Friant-Kern Canals to the north and south of Millerton Lake near the headwaters of the San Joaquin River. USBR also executed a separate Exchange Contract [48] in 1939 for 840,000 AF of cropland water rights that had originally belonged to Miller & Lux, but which had since been bequeathed to the four private canal companies that were created as Miller & Lux was waning: Central California Irrigation District, Columbia Canal Company, San Luis Canal Company, and Firebaugh Canal Company. The Exchange Contract lays out the substitution agreement under which these users agreed to forego diversion of San Joaquin River water in exchange for substitute supplies from the Sacramento River stored at Shasta Dam. Importantly, the agreement did not require the Exchange Contractors to relinquish possession of their water rights.

According to the contract, in a "typical" year, the Exchange Contractors are to receive 100% allocation (840,000 AF). In a critical dry year, the substitute Sacramento River water deliveries can be reduced to 650,000 AF (about 78% of the maximum allocation) [48]. In the event of a shortage severe enough that USBR is unable to send the exchange contractors at least that amount from the Sacramento River, the Exchange Contractors may request that the deficit be made up with water from their original source, which is stored in Millerton Lake. When this happens, CVP contractors with more junior rights may lose access to water that would otherwise come to them. This is how the priority system can come into play, and why it is possible for the Exchange Contractors to receive most of their contracted water while many other irrigators go completely without.

Stroshane [47] (p. 16) likens the purchase and exchange contracts to the way the U.S. Constitution relates to the union of states: "[t]he contracts provide a framework and a point of departure for water project operations every single year. Their effect is nothing if not constitutional and foundational for the Central Valley Project." The Exchange Contract enshrines the priority system and specifically the old Miller & Lux monopoly's control over San Joaquin River basin land and water in perpetuity.

3.2. Sacramento River Settlement Contracts

The Purchase and Exchange Contracts made it possible to distribute water stored at Millerton Lake to contractors in the Friant Division and along the Madera Canal, but USBR still had a major problem, which was that they had no way to protect "their" water in the Sacramento River from being intercepted by the several hundred diverters along the mainstem after it was released from Shasta Reservoir but before it could reach the pumps in the Delta that move the water to the Exchange Contractors.

Again, rather than pursue an adjudication of Sacramento River rights, USBR opted to negotiate some 145 Settlement Contracts, altogether totaling some 2.2 million AF of face water [49]. As with the Exchange Contract, these contracts preserve seniority. They divide up water allocation between "base supply" and "Project water." Base supply is the amount of water the contractor is allowed to divert for free, in deference to these users' senior rights. Project water is an amount over and above the base supply that can be purchased from USBR. Glen-Colusa Irrigation District, for example, is allowed to divert 720,000 AF a year of base supply for free and also are entitled to purchase 105,000 additional AF of "Project water," for a total of 825,000 AF per year, doled out in monthly maximums specified in the contract [50].

The Settlement Contracts have the same critical year trigger criteria as the Exchange Contract but the shortage provisions are different. The Settlement Contracts simply state that in a critical year, their total water will be reduced by up to 25%. This reduction is assessed monthly from April to October.

3.3. Water Service Contracts

At the bottom of the CVP hierarchy are the Friant Division contractors and those south and north of the Sacramento/San Joaquin Delta with regular water service contracts. Water service contracts comprise 16% of the total contracted volume of water in the North-of-Delta (NOD) CVP users, dwarfed in volume by the Settlement Contractors (Figure 2). In the South-of-Delta (SOD) section, the water service contracts comprise 71% of the SOD grand total, with the Exchange Contractors having rights to 28% of the total contracted SOD volume. SOD group of water service contracts includes the Westlands Water District, the biggest irrigation district in the U.S.

Water service contracts may be for agricultural or municipal and industrial (M&I) uses. There are clear rules about what happens to the M&I contracts in the event of a "condition of shortage" [51]. When a contractor has both agricultural and M&I uses, irrigation has to decrease by 25% before any M&I reductions are required.

The Friant Division of the CVP includes irrigators withdrawing water from either the Madera or Friant-Kern Canals, stored at Millerton Lake. The Friant Division's contracts were started in 1951 when Friant Dam and Madera Canal were finished and have a term of 40 years [28] (p. 779). Their water is divided into Class 1 and Class 2, the former being more reliably available than the latter.

Most of these contractors receive water from the Friant-Kern Canal; the Madera group is just two districts, Chowchilla Water District and Madera Irrigation District. However, these two districts have contracts for 18% of the total Friant Division Class 1 water. The other 83% is distributed among some 30 other entities with contracts for water for M&I, agriculture, or both. Most of the water either belonged or still technically belongs to the four Exchange Contractors.

Allocations for the Friant Division are based on a formula specified in their contracts (see, for example, the water service contract for Tulare Irrigation District [52]). The maximum contracted volume of Class 1 water is reduced by the ratio of actually available water divided by the total claims for that water. In times of shortage, however, the junior status of the Friant Division and the NOD and SOD water service contractors means they can be reduced even to 0%.

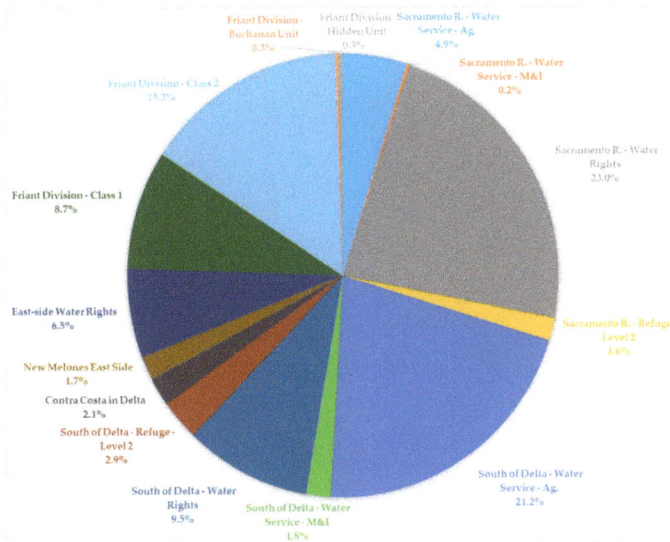

Figure 2. The proportional distribution of CVP's 9.5 MAF of total maximum contracted water supplies among the different user groups. Percentages derived from data reported by U.S. Bureau of Reclamation for 2016 [53].

4. Impacts and Outcomes for CVP Contractors during the 2012–2016 Drought

In their 2016 Drought Contingency Plan for the CVP and State Water Project (SWP), USBR and California Department of Water Resources took stock of the unprecedented nature of the drought up to that point:

> *"California has just ended its fourth consecutive year of below-average rainfall and snowpack, and Water Year (WY) 2015 was the eighth of nine years with below-average runoff. This extended drought has produced chronic and significant shortages to municipal and industrial, environmental, agricultural, and wildlife refuge water supplies and led to historically low groundwater levels. This recent dry hydrology has set many new statewide records, including the driest four-year period of statewide precipitation (2012–2015). In calendar year 2013, many communities recorded their lowest-ever levels of annual precipitation; calendar year 2014 saw record-low water allocations for the Central Valley Project (CVP) and State Water Project (SWP) contractors; and January 2015 was the driest January on record for precipitation Statewide. WY 2015 also produced by far the lowest snowpack in the Sierra Nevada since records have been kept, and by some estimates based on tree-ring analysis, was the lowest over the past five centuries".* [54] (p. 4)

In this section, I unpack how these impacts were distributed through the CVP's user groups and how they were mediated by the underlying system of water rights and contracts.

4.1. Allocations and Curtailments for Central Valley Project Water Users, 2012–2016

Although I focus here on CVP allocations, it should be noted that thousands of water rights were simultaneously curtailed by the California State Water Resources Control Board (SWRCB) based on §1058.5 of the California Water Code which allows SWRCB to implement temporary emergency water regulations [55]. This is noteworthy because these senior rights had not been curtailed since the 1976–1977 drought [56].

The hierarchy of state water rights embedded in the CVP contracts is evident from the allocations made to various user groups during the drought years. Table 1 presents water allocations to the various CVP water user groups by Project division during the drought years 2012–2016 based on data reported by USBR [6].

Table 1. Drought year allocations to CVP water user groups by region. Allocations are reported as percentages supplied out of total contracted volumes. Allocations are averaged for February–May and June–August except for 2015, in which no changes to the initial February 27 allocation were reported.

CVP Region	Water User Group	2012		2013		2014		2015		2016	
		Feb.–May	June–Aug.	Feb.–May	June–Aug.	Feb.–May	June–Aug.	Feb.–May	June–Aug.	Feb.–May	June–Aug.
North of Delta	Agricultural water service contractors	77%	100%	75%	75%	0%	0%	0%	0%	100%	100%
	Urban contractors (M&I)	92%	100%	100%	100%	50%	50%	25%	25%	100%	100%
	Wildlife refuges	92%	100%	100%	100%	75%	75%	75%	75%	100%	100%
	Settlement Contractors/Water Rights	92%	100%	100%	100%	63%	75%	75%	75%	100%	100%
	American River contractors (M&I)	*	*	75%	75%	50%	50%	25%	25%	100%	100%
	Contra Costa	*	*	75%	75%	50%	50%	25%	25%	100%	100%
South of Delta	Agricultural water service contractors	37%	40%	21%	20%	0%	0%	0%	0%	5%	5%
	Urban contractors (M&I)	75%	75%	71%	70%	50%	50%	25%	25%	55%	55%
	Wildlife refuges	92%	100%	100%	100%	48%	65%	75%	75%	100%	100%
	Settlement Contractors/Water Rights	92%	100%	100%	100%	48%	65%	75%	75%	100%	100%
Friant Division	Friant (Class 1 water)	43%	50%	56%	56%	0%	0%	0%	0%	40%	70%
Eastside Division Contractors		100%	100%	100%	100%	55%	55%	0%	0%	0%	*

Allocation color code:

75–100%	50–74%	25–49%	0–24%

* No value reported.

In the first two years of the drought, the Settlement and Exchange Contractors received full or nearly full allocations. In 2012, the first official year of the drought, all NOD users received full allocations while South of the Delta, agricultural service contracts were reduced to 40%; the Friant Division received 50% of Class 1 water and zero Class 2 water; and the Exchange Contractors received full allocations.

The next year, NOD agricultural water service contracts were curtailed to leave water for urban uses, wildlife refuges, and the Settlement Contractors, which all received full allocations. The agricultural water service contractors were reduced by 25% for all of 2013. SOD agricultural contractors were also cut back to around 20% for all of 2013. Friant Division curtailments varied from 45–65% from early late February to mid-July, with an average of 56% allocation, while the Exchange Contractors received 100% allocations.

2014 is notable for being the first time in history that SOD senior rights holders received allocations less than 75%. When USBR was unable to provide the Exchange Contractors with the minimum contracted amount from the Sacramento River, the Contractors exercised their rights to the San Joaquin River. With the exception of 2014, the Exchange Contractors received at least 75% allocations in every other year of the drought. Both the Friant Division and SOD agricultural water service contractors received no water two years in a row (2014–2015), and the latter group only received a 5% allocation in 2016 even as hydrologic conditions improved relative to the previous years. The NOD agricultural water service contractors also received 0% allocations during 2014–2015 while the Sacramento River Settlement Contractors generally received their contracted 25% maximum reductions.

In 2016, hydrologic conditions improved in the Sacramento Valley and starting April 1 all NOD users received full allocations. However, conditions remained severe for most SOD users who received only 5% for agricultural water service contracts, 55% for M&I, and 30% for the Friant Division. In reaction to this disparity between north and south, the director of the California Farm Water Coalition said in spring 2016: "you've got Lake Shasta at 90 percent capacity and probably going to be full by the summer, and Folsom reservoir filling up, and south of the Delta farmers are getting almost no water–it's a screaming headline that the system is broken" [11], a sentiment echoed by a representative of the Westlands Irrigation District [57].

These disproportionate allocations and impacts are not because Friant Division has much larger acreage and thus a much larger water allocation in a maximum year. On a per-acre basis, senior water rights holders' maximum allocations are much larger than the Friant Division. Settlement Contractors are entitled to about 4.7 AF/ac (2.12 million AF for 450,000 acres) and the Exchange Contractors 3.5 AF/ac (840,000 AF for about 240,000 acres), while the Friant Division contracts are for between 1.9 and 0.71 AF/ac depending on whether the unreliable Class 2 water is counted (much of which is used for groundwater replenishment rather than field crops). In other words, the baseline of maximum allocations is uneven. As explained earlier, this is related to the relatively late addition of these acres in time relative to the senior rights holders elsewhere in the San Joaquin Valley.

4.2. Socioeconomic Impacts of CVP Curtailments

The California agricultural economy survived the drought in relatively good shape overall [58], but not without serious costs. Estimates of the total direct costs drought-induced water shortages on the statewide agricultural economy ranged from an estimated $1.5 billion in 2014 to $1.8 billion in 2015, to $550 million in 2016 [2,4,59].

However, the statewide metrics mask the geographical unevenness of the negative impacts of the drought. The worst impacts among CVP contractors were visited most heavily upon people in certain parts of the San Joaquin Valley (Table 1) [3,4]. In each of the years 2014, 2015, and 2016, among the three CVP basins (Sacramento, San Joaquin, and Tulare Lake) the 0% allocations had the greatest effect in the Tulare Basin where the Friant Division is located. The effects of fallowing and lost crop revenues were most severe in that region, where surface water losses were 3 million AF [59]. In 2014, fallowing was somewhat evenly distributed across the basins, with the Sacramento basin having the most at

151,000 acres fallowed compared to average; Tulare Lake Basin growers removed 133,000 acres from production. However, in the Tulare Lake Basin where fruit and nut trees are prevalent, crop revenue losses were greater than in the Sacramento Valley, Delta, and East of Delta regions at $373,400,000 [59].

The disparities increased in 2015; over half of all the fallowed acres in the Central Valley were in the Tulare Lake Basin (an estimated 80,000 acres relative to average water supply conditions), along with 67% of the total crop revenue losses for the entire state [2]. This is less than 1% of all irrigated land in California, but about 90% of the drought-induced fallowing is in the Central Valley south of the Delta. In 2016, 100% of the fallowing in the Central Valley occurred in the Tulare Lake Basin [4].

Fallowing is associated with unemployment and underemployment. Agricultural job losses in 2014 were estimated to be over 17,000, significantly greater than in the 2009 drought; in fact, 2014 drought impacts were estimated to be up to 50% more severe than 2009, which has been attributed to the much lower water supplied by CVP and the SWP, notably the Friant Division's 0% allocation [59]. In 2015 there were more job losses in the Sacramento Basin than Tulare Basin (5480 compared to 3850). In 2016 an estimated 1815 full and part-time agriculture-related jobs were lost in the Central Valley (4700 when indirect effects on other sectors are included), with about 75% in the Tulare Basin. Medellin et al. (2016) reported that 57% of the $6.5 million in direct assistance provided through the state's Drought Emergency Assistance Program (DEAP) went to farm-related workers in the San Joaquin and Tulare basins in 2016. This is unfortunate given that it is already one of the poorest regions of the state. Kern, Tulare, Kings, and Fresno Counties (as well as Merced and Madera) all have poverty rates between 20–25% [60].

4.3. Groundwater Impacts of CVP Curtailments

Surface water deficits experienced by some irrigators were largely made up for by increasing groundwater pumping [61]. Groundwater substitution appears to have been greatest in South of Delta regions (Tulare Lake Basin in particular) with the greatest surface water deficits to make up as a result of severe surface water curtailments. In fact, new wells proliferated during the drought in areas with major surface water reductions; over 5000 were drilled in the San Joaquin Valley between 2012–2015, more than in the previous 12 years combined [62]. In 2015, the most wells went in in Tulare county (904) and Fresno county (627), followed by Merced county with 304 [62].

Negative impacts of groundwater overdraft include increased pumping costs, land subsidence, and permanent loss of storage capacity. Exacerbating declines in local water tables may contribute to dry domestic wells in marginalized unincorporated communities such as East Porterville and Woodville [62].

4.4. Conflict over CVP Operations during the Drought

The 2012–2016 drought is notable for the ways in which distribution of CVP water generated informal tensions between senior and junior growers within the system [63]. One formal conflict stemmed from the release of water from Millerton Lake in May 2014 to fill the rights of the Exchange Contractors when USBR could not provide them with their contracted minimum substitute supply from the Sacramento River [64]. This occurred while Friant Division received 0% allocations.

In May 2014, Friant Water Authority led 21 plaintiffs in legal action to stop USBR from releasing water from Friant Dam to wildlife refuges and the Exchange Contractors but were denied [65]. The plaintiffs contended that in allocating water to the Exchange Contractors and wildlife refuges but not to them, USBR breached their water service contacts, in addition to taking property (water rights) without compensation in violation of the 5th Amendment of the U.S. Constitution [66]. Intervening in the case on the side of USBR were not just the refuge management organization, but also the San Joaquin River Exchange Contractors Water Authority, the San Luis and Delta-Mendota Water Authority, and the Westlands Water District [67].

After this, the case was dropped [67] but re-filed in 2016 [68]. The revised complaint did not specifically question the USBR's interpretation of the Exchange Contract; rather, it requested $350 million in compensation for the alleged taking of their property under the 5th Amendment [69].

How did the Friant Division respond to the outcomes of the drought? The Friant Water Authority identified protection of water rights and agreements and the development of a sustainable water supply as their primary goal [70]. The protection of existing water rights and supplies included the resolution of " ... issues caused by the Bureau of Reclamation's (Reclamation) interpretation of the Exchange Contract" which the Authority listed as one of several threats to their water supply, particularly to their ability to ensure some minimum surface water allocation during drought periods [70] (p. 4). The legal conflict and internal response by Friant Water Authority is evidence of the drought and the priority system embedded in the contracts causing tension between not just environmentalists and agriculture, but between different groups of irrigation communities.

5. Discussion

Based on the foregoing analysis, we can see the important role that water rights play in dictating uneven water allocations and their ripple effects. While water rights are not the only factor affecting allocations, they are an especially important one during shortages. Because there is a definite hierarchy based on chronological order of rights and contracts acquisition strongly affecting CVP drought allocations, it is worth revisiting earlier critiques of the equity dimensions of prior appropriation before considering a more equitable alternative.

5.1. Fairness Critiques of Priority in Relation to the CVP Drought Allocations

On its face, priority (in time) is not an unfair basis for allocating water resources. It arguably provided a greater degree of certainty to the individual making risky investments of money and labor to irrigate the desert than did the correlative rights of the riparian doctrine. As water law scholar Samuel Wiel wrote, "the pioneer gambles for high stakes, and unless those large stakes are secure to him after he succeeds, he will not gamble, and there would be no pioneering" [71] (p. 534). Dunning [38] reminds us that the priority principle was described as a "maxim of equity" in *Irwin v. Phillips* (1855), the influential early California case of competing appropriation and riparian rights claims.

Today, even junior rights holders see the priority system as a generally fair one. For example, when the Friant Water Authority took action to block USBR from sending water from Millerton Lake to the Exchange Contractors and wildlife refuges in May 2014, an Authority spokesperson was careful to clarify that they were not attacking the priority system itself or the Exchange Contractors, but rather USBR's decision-making [66].

However, critiques of California's water rights system proliferated during the drought [12–15,17,72], echoing a long history of critiques of the fairness of the priority system of allocation and curtailment. Late 18th and early 19th century critics of reliance on the relatively narrow criterion of chronology of rights establishment for determining winners and losers included Elwood Mead, Frederick Newell, John Wesley Powell, as well as a number of western judges, scientists, and engineers [24,73]. A minority of 20th century scholars have carried forward these earlier critiques of the equity and fairness aspects of seniority as a water allocation scheme (e.g., [73–75]).

Samuel Wiel argued that the fairness of the priority system was conditional on whether a basin was (a) not fully appropriated and there was considerable risk to a pioneering individual in attempting irrigation development versus; (b) a system long after the pioneer times which is fully appropriated [71]. In the case of the latter, he argued, curtailments cannot be distributed without gross disparities based on historical happenstances of priority dates.

The CVP and the rivers it draws from are fully appropriated and have been for decades, yet shortages are managed largely based on deference to 19th-century water rights. Even though the California water rights system contains riparian rights, the evidence presented above from the

2012–2016 drought illustrates how the priority system underpins CVP allocations during drought, supporting critiques of the fairness of seniority systems in fully allocated surface water bodies.

The critique of priority can be extended by drawing from Ingram et al.'s five principles to consider how the evidence presented may amount to an inequitable situation. Like the Colorado River Compact, the bedrock of CVP system operations is a complex set of different promises (water rights and permits granted, agreements entered, contracts signed) which must be respected. However, if equity is understood as a balancing exercise, the effects of enforcing these promises must be weighed against other criteria. If satisfying one of the five principles causes an imbalance in the other constituent principles, then it is safe to conclude that equity is not being satisfied.

In this case, one of the bigger balancing challenges is weighing the principles of honoring past promises with reciprocity (a condition of fair sharing of distributive advantages and costs by all members of the relevant community). The evidence presented above suggests that, overall, the drought curtailment system is uneven enough as to fail this test. One element of reciprocity is that those who use more water should expect to have to sacrifice more under conditions of scarcity [40]. In the San Joaquin basin, the Friant Division's maximum allocation of Class 1 water is less than the Exchange Contractors' total water rights and on a per-acre basis is comparatively almost half as much, yet during severe drought the Contractors are entitled to receive San Joaquin water to satisfy their demands before the Friant growers receive anything. As discussed above, the USBR's enforcement of this rule was a point of formal conflict during the drought.

This raises a key question: at what point do conditions become sufficiently different from those that existed when promises and contracts were originally made as to warrant renegotiation? In line with Wiel's argument, Ingram et al. [40] note that negative effects are generated when the priority system is pushed to an extreme. In this case, the serious negative social and ecologic impacts that resulted during the 2012–2016 drought stemmed in large part from the conditions Wiel described almost a century ago (full allocation of surface water). This situation, combined with robust predictions about the hydrologic impacts of climate change on the Sacramento and San Joaquin Rivers, amounts to a sufficiently changed set of conditions relative to the mid-20th century as to justify a reordering of water right priorities. However, any reordering should be done in balance with the other four equity principles.

5.2. The Connection between CVP Allocations and the Intergenerational Inequity of Long-Term Central Valley Groundwater Depletion

This analysis has focused so far on the fairness of drought allocations and their impacts, but this is only one of potentially numerous factors that may be relevant to overhauling water rights and allocations for CVP users. A small sample might include such things as the degradation of ecosystems and extirpation of endemic species; injustices to indigenous tribes; the widely disproportionate subsidies and costs of water between agricultural and M&I users; even the fraudulent acquisition of land and water rights under both riparian and appropriation doctrines during the laissez faire era by Miller & Lux and others.

In the case of the CVP, one of the important ripple effects of 5% or less water allocations is increased groundwater pumping. Increasing groundwater pumping to make up for a lack of surface water during a drought is a form of conjunctive use that can, under certain conditions, be a valuable adaptive strategy to socioecological disturbances [76]. In the San Joaquin Valley, however, where natural replenishment rates are often very slow and regional groundwater basins have been in continual decline for nearly a century, using groundwater as a drought buffer exacerbates a longstanding problem [61], further inhibiting the realization of the principle of intergenerational equity.

There is an important historical dimension to this problem. While drought impacts are most severe in the Tulare Lake Basin, it is also true that groundwater depletion has continued overall and has been concentrated in the same area [77]. Growers in the region who contracted for CVP supplies in some cases used them to develop new lands instead of to replace groundwater pumping for watering existing acreage as was intended by the Project's proponents [74] (pp. 277–278). All of the basins in the area have been designated as being in critical overdraft since 1980, yet cumulative groundwater depletion of the Central Valley aquifer has only increased as depletion has accelerated [61] and storage capacity has decreased [78,79]. The additional storage created by the CVP and SWP ended up doing relatively little to address the problem, which has only gotten worse with time.

The benefits of groundwater pumping during drought are mostly individual and short-term but the aggregate costs are mostly socially distributed and deferred to future generations in the form of permanently reduced aquifer storage capacity, land subsidence, and increased costs of pumping from chasing the water table further and further down.

Adaptive solutions for the Central Valley overdraft problem are somewhat limited. Recharge can be increased through water storage, but major new reservoirs are not usually considered politically realistic. Depletion can be slowed by increasing irrigation efficiency instead of continuing to flood irrigate, but this is also a problematic option because (a) inefficient irrigation replenishes streams and aquifers to some extent and (b) saved water is often just used to irrigate more acreage or more high value but water-intensive crops (as has happened with the proliferation of almonds and other thirsty nut trees in parts of the Tulare Basin) [77]. Aquifer recharge by irrigation districts when floodwater is available to capture may mitigate the overdraft problem to some extent in certain areas, but on its own it seems unlikely to bring critically overdrafted basins into some state of long-term sustainable yield without significant concomitant reductions in pumping.

Given this limited set of options, maintaining the status quo of rights and contracts in which 0% surface allocations are highly likely for the most groundwater-reliant CVP customers seems likely to generate the same outcomes over and over, worsening the overdraft problem. This raises considerable questions about the ability of communities of junior CVP contractors in the San Joaquin Valley to adapt to a changed climate characterized by reduced snowpack and runoff and more frequent and severe drought episodes. The supply problem may only become more difficult if Groundwater Sustainability Agencies pass meaningful pumping constraints under the Sustainable Groundwater Management Act of 2014 which reduce growers' ability to continually rely on groundwater as a drought buffer when they are cut off from surface supplies.

This raises some very difficult choices. The main options are to (1) reallocate some surface water from senior users such as the Exchange Contractors to more evenly distribute the water that is available in Millerton Lake during drought in order to help mitigate groundwater overdraft in the Central Valley, or (2) substantially shrink the overall water footprint of agriculture in the San Joaquin Valley in general to reduce the total demand for both groundwater and CVP water. With regard to option (1), what degree of responsibility for resolving the problem can be placed on CVP contractors in the San Joaquin Valley who used their surface water supplies to irrigate new land instead of using it to reduce groundwater overdraft as was the original intent of the Project? This is also a potential point of conflict, since addressing the groundwater problem may require making more surface water available for recharge, at least some of which would likely come from the senior entitlements of other users.

At a minimum, in order to rebalance equity principles so that intergenerational equity is more properly weighted, a 50–100 year planning horizon must be adopted [42]. Depending on how it is implemented, the requirements of the 2014 Sustainable Groundwater Management Act may be instrumental in eventually achieving such a long-term vision. Additionally, long-term declines would have to be halted and eventually reversed through natural and artificial recharge. This would involve more strictly operating the Friant Division as the conjunctive use project it was intended to be, such that temporary withdrawals from groundwater reserves in dry years are later replenished instead of contributing to continued net depletion.

I do not propose to answer the difficult questions raised by the need for equitable adaptation to the hydrologic impacts of climate change. Instead, I argue that surface water allocations cannot be understood as disconnected from groundwater usage, and thus any reordering of surface water priorities within the CVP universe of contractors to reduce distributional inequities of socioeconomic suffering during drought must be balanced with the need to address the intergenerational inequity of long-term groundwater mining.

As Gleeson et al. [41] (p. 379) point out, "the social and economic benefits of large aquifer withdrawals may not compensate for the significant depletion of aquifers that are effectively non-renewable on human timescales" (this applies also to alluvial unconfined aquifers and groundwater quality [42]). The capital generated by industrialized agriculture in the Central Valley cannot alone be used to justify current patterns of groundwater extraction on equity grounds if those patterns substantially reduce the capacity of future generations to adapt to a climate with more frequent and severe drought episodes and shrinking snowpack.

5.3. Equitable Apportionment as a Fairer Alternative to Priority

What legal doctrine might one apply if the system could be torn down and rebuilt in order to strike a better balance between the five equity principles? I suggest equitable apportionment is one hypothetical alternative to the current system which is more conducive than priority to a broadly fair and sustainable water management regime.

The doctrine of equitable apportionment dates back to the first interstate surface water dispute that resulted in *Kansas v. Colorado* (1907) and has since developed primarily within that subcategory of water resource law [28,80]. In its decision for *Nebraska v. Wyoming* in 1945, the U.S. Supreme Court affirmed that "[e]quitable apportionment among appropriation States does not require a literal application of the priority rule" [81]. Further, the court identified several categories of relevant criteria beyond priority, including

> " . . . physical and climatic conditions; the consumptive use of water in the several sections of the river; the character and rate of return flows; the extent of established uses; the availability of storage water; the practical effect of wasteful uses on downstream areas; the damage to upstream areas as compared to the benefits to downstream areas if a limitation is imposed on the former. The decree of equitable apportionment to be entered in this case must deal with conditions *as they exist at present and must be based on the dependable flow of the river* which is not greater than the average condition which has prevailed since 1930". [81] (emphasis added)

Although the development of equitable apportionment has happened primarily within interstate water law, different forms of it have been applied at times in California for both groundwater and surface water. For example, Dunning [38] has argued that under a strict priority rule, SWRCB should have imposed Sacramento-San Joaquin Bay-Delta salinity control requirements disproportionately on the junior appropriator, which would have placed the greater or perhaps sole burden of maintaining outflows for salt flushing on the SWP and little or no responsibility on the CVP. Instead, a sharing rule was adopted which " . . . reflects a form of state equitable apportionment of water resources [38] (p. 109)," a fairer arrangement in light of the total body of relevant facts and evidence than priority would have dictated.

From an equity standpoint, then, one advantage equitable apportionment has over priority is that it allows for a variety of non-priority criteria to be brought to bear on court decisions. Ruhl [82] (p. 52) concluded that " . . . equitable apportionment encompasses whatever seems relevant to a fair division of the resource between the states. This means equitable apportionment is a flexible doctrine, able to incorporate new knowledge not only about water demands and uses, but also about the ecology of water in general."

Further, because equitable apportionment allows for priority to be used as a guiding principle, it does not require an *equal* apportionment. This is another reason why it would be appropriate for overhauling CVP allocations and shortage provisions, which I submit are inequitable in certain ways but not *unjust*. One key difference between justice and equity is that they correspond with different kinds of remedies. In some cases equality is what is needed, e.g., uniform national water quality standards and equal protection from water-related hazards for all economic and ethnic groups [27] (p. 80). Other cases such as fair access to water resources, compensation for injury, a minimum supply, and procedural justice, may be more suited to an equity-driven approach that "may secure remedies that are deemed fair by many but not equal for all" [27] (p. 76). This is the appropriate kind of remedy needed in the case of the CVP.

5.4. Anticipated Opposition to Water Rights Reform for the CVP

In the case of the CVP, re-tooling the system to achieve a more even balance of equity principles would likely entail revising water supply contracts and some reordering of priorities, along with establishing a new shortage sharing system that goes beyond strict priority and contracts. At minimum, all rights, including the pre-1914 unregulated rights, would need to be quantified [83]. Opposition to quantification and regulation dates back to Miller & Lux's fights against it in the courts in the 1800s and would likely continue to be opposed by the holders of the most senior rights such as the Exchange Contractors. To the extent that an overhaul (whether equitable apportionment by the courts or some other way) would result in the reordering of priorities, reform can be expected to be opposed primarily by those who benefit most from the status quo: senior rights holders who suffer the least from drought and strongly support strict adherence by SWRCB to strict enforcement of the priority system [84]. Northern growers more generally can be expected to resist any rules that provide more exports from the north to the Friant Division and the junior SOD water service contractors, especially if perceived as a bailout of southern growers who rely heavily on mined groundwater.

Any overhaul of water rights involving modifications to, and reordering of, priorities can be expected to generate conflict. This is not inherently a bad thing. However, it is also true that the current priority system and its implementation by SWRCB is already plagued with conflict, which proliferated during 2012–2016. I noted above just one instance, which pertained specifically to the CVP. However, it is worth stating that a water governance system stymied by conflict cannot be an effective one, nor can it be considered a well-adapted one. Without a larger political solution that resolves problems of fairness, there is little reason to expect that future droughts will be markedly less conflict-laden than the most recent one.

5.5. Equitable (Re)Apportionment Implementation Scenarios

This raises an important question: what kind of process might be used to apply equitable apportionment to generate fairer drought outcomes? From an equity perspective, who applies equity principles and how is very important. In one hypothetical scenario, a legal dispute could lead to a court-supervised adjudication using equitable apportionment, as has been done in a number of California groundwater basins. There is some indication from past decisions that the state courts would not likely reject an equitable (re)apportionment of surface water rights as long as priority is not completely ignored [38,85].

Some potential drawbacks associated with a judicial reapportionment are: the risk that the courts could construe equity too narrowly; the transaction costs to stakeholders of time and legal fees could be very great, especially if the legal proceedings last many years; significant costs may greatly disadvantage the less powerful and/or well-heeled third-party interests most affected by past and existing inequities.

A second, more preferable scenario might be a legislative one involving a statutory reapportionment administered by the SWRCB according to its authority to alter and condition water rights and permits on behalf of the public trust [86]. Brian Gray [87] (p. 237) has argued that Section 2 of Article X of the California Constitution [88] " . . . confers broad authority on the state to modify existing water rights to ensure that the current apportionment of California's water resources serves contemporary economic, social, and environmental goals in a reasonably efficient manner." SWRCB has at times exercised its authority under Section 2 and the Water Code to condition the exercise of water rights and permits in response to environmental problems without following priority [38].

Procedurally, an equitable reapportionment initiated and administered by SWRCB could take any number of different forms. Unilateral action by SWRCB would likely result in a maximum of political and legal conflict and be unlikely to generate a new arrangement that is accepted by the affected communities of users if the process deprives them of collective agency. The equity principles of value pluralism and the accommodation of multiple value claims in resource allocation decision processes suggest a need for some amount of collective, participatory action so that the values of water rights holders and third parties can be meaningfully included. A messier but more equitable process than unilateral action could entail some type of sustained, mediated negotiations overseen by SWRCB involving a broadly inclusive community of rights holders and third parties to articulate what an equitable basic apportionment and a set of drought curtailment provisions would look like. These could then be implemented in the form of revised permits and contracts by SWRCB and USBR. Besides satisfying procedural equity criteria, another benefit is the possibility of identifying nontraditional ways of responding to drought that increase both equity and adaptability such as voluntary conservation arrangements [89].

However, if Hanemann and Dyckman [90] are correct in their analysis of CALFED's failures, purely voluntary negotiations would be likely to fail without strong oversight and direction from the state, given that many of the same interest groups involved in CALFED would be involved in an equitable apportionment for the CVP. In such a scenario, it would be important for legislators and regulators to build from the positive procedural aspects of past stakeholder-driven efforts while avoiding their critical flaws such as the marginalization of environmental justice interests [91] and the lack of regulatory power and a hard statutory requirement to achieve a final resolution that hampered CALFED [90,92–94]. California could perhaps look to more successful Endangered Species Act-driven stakeholder negotiation processes in other states for ideas. For example, the Edwards Aquifer Recovery Implementation Program in Texas demonstrated the importance of designing authorizing legislation with clear deadlines and meaningful consequences for missing them for keeping all parties at the table even when they are far apart on a particularly thorny issue [95].

6. Conclusions

As Stroshane [47] (p. 168) has stated, "drought reveals the tensions in governing a capitalist, hydraulic society such as California, where competition over scarce water supplies drives conflict." A central vexing question for the state is how to deal with those tensions, knowing that there will be winners and losers with each new round of drought. Like the Colorado River Compact, the CVP is at bottom a set of promises in the form of its Settlement, Exchange, and water service contracts. The promises were always unequal; some were promised greater security in the event of shortages than others. However, also like the Colorado River and numerous other surface water bodies in the West, the resource was over-promised from the beginning. USBR delivers on average far less water than is contracted for.

In this analysis I have explored how these promises in the form of the priority system of water rights underlying the CVP water supply contracts becomes highly important during drought in determining the relative winners and losers of water distribution. I have demonstrated how this system generates uneven socioeconomic outcomes for different communities of water users. I have argued that the outcomes of CVP's contracts and shortage provisions during the drought lend support to critiques of the fairness of the priority system, and that equitable apportionment is an example of how a complex water allocation and management system like the CVP could be overhauled to incorporate and advance equity principles that currently are out of balance. Last, I have emphasized the importance of intergenerational fairness (or lack thereof) of the Central Valley overdraft problem, which, via its connection to CVP allocations, is worsened during drought.

As California and other Western states do the difficult work of increasing capacity to adapt to major climatic disruptions such as drought, it is important to do so in ways that increase equity among humans and non-humans and avoid exacerbating existing environmental and socioeconomic inequities. This will likely require a complex balancing process that may involve state management and regulatory agencies, federal agencies, and state and federal courts, as well as communities of users and third-party interests. Therefore, strong political leadership will be critical.

Acknowledgments: I wish to thank the editors and two anonymous reviewers for helping me improve this manuscript. I also thank the Martin Daniel Gould Center for Conflict Resolution at Stanford Law School for the generous assistance with publication costs.

Conflicts of Interest: The authors declare no conflict of interest.

References

1. Griffin, D.; Anchukaitis, K.J. How unusual is the 2012–2014 California drought? *Geophys. Res. Lett.* **2014**, *41*, 9017–9023. [CrossRef]

2. Howitt, R.; MacEwan, D.; Medellin-Azuara, J.; Lund, J.; Sumner, D. *Economic Analysis of the 2015 Drought for California Agriculture*; Center for Watershed Sciences: Davis, CA, USA, 2015; p. 16.

3. Castellon, D. Valley agriculture hit hardest by the drought. *Visalia Times-Delta.* 15 July 2014. Available online: http://www.visaliatimesdelta.com/story/news/local/2014/07/16/valley-agriculture-hit-hardest-drought/12719953/ (accessed on 15 July 2017).

4. Medellin-Azuara, J.; MacEwan, D.; Howitt, R.E.; Sumner, D.A.; Lund, J.R. *Economic Analysis of the 2016 California Drought on Agriculture*; Center for Watershed Sciences: Davis, CA, USA, 2016; p. 17.

5. Rodriguez, R. Valley's Westside farmers seethe over tiny water allocation from feds. *Fresno Bee.* 1 April 2016. Available online: http://www.fresnobee.com/news/local/water-and-drought/article69443782.html (accessed on 1 August 2017).

6. U.S. Bureau of Reclamation Central Valley Operations Office. *U.S. Bureau of Reclamation Summary of Water Supply Allocations*; U.S. Bureau of Reclamation: Washington, DC, USA, 2017. Available online: https://www.usbr.gov/mp/cvo/vungvari/water_allocations_historical.pdf (accessed on 17 November 2017).

7. U.S. Bureau of Reclamation. *Long-Term Central Valley Project Operations Criteria and Plan*; U.S. Department of the Interior: Sacramento, CA, USA, 2004; p. 238.

8. U.S. Bureau of Reclamation about the Central Valley Project. Available online: https://www.usbr.gov/mp/cvp/about-cvp.html (accessed on 25 November 2017).

9. Brougher, C. *California Water Law and Related Legal Authority Affecting the Sacramento-San Joaquin Delta*; Congressional Research Service: Washington, DC, USA, 2008; p. 11.

10. Cody, B.A.; Folger, P.; Brown, C. *California Drought: Hydrological and Regulatory Water Supply Issues*; Congressional Research Service: Washington, DC, USA, 2015; p. 42.

11. Bland, A. Why Some Farmers Get Little Water Despite Rain. *Water Deeply.* 15 April 2016. Available online: https://www.newsdeeply.com/water/articles/2016/04/15/why-some-farmers-get-little-water-despite-rain (accessed on 18 November 2017).

12. Kahn, D. Calif.'s quirky water rights system is showing its age. *EE News.* 25 June 2015. Available online: https://www.eenews.net/stories/1060020893 (accessed on 8 November 2017).

13. Los Angeles Daily News Editorial Board California is drowning in ancient and unfair water rules: Editorial. *Los Angeles Daily News.* 21 November 2013. Available online: http://www.dailynews.com/2013/11/21/california-is-drowning-in-ancient-and-unfair-water-rules-editorial/ (accessed on 9 November 2017).

14. Los Angeles Daily News Editorial Board It's time to reform California's inherited water rights: Editorial. *Los Angeles Daily News.* 8 May 2015. Available online: http://www.dailynews.com/opinion/20150508/its-time-to-reform-californias-inherited-water-rights-editorial/1 (accessed on 8 November 2017).

15. Morin, M. As California drought worsens, experts urge water reforms. *Los Angeles Times.* 1 June 2015. Available online: http://www.latimes.com/local/lanow/la-me-ln-water-rights-20150601-story.html (accessed 8 November 2017).

16. Parrish, W. A Solution for California's Water Woes. *East Bay Express.* 15 July 2015. Available online: https://www.eastbayexpress.com/oakland/a-solution-for-californias-water-woes/Content?oid=4412629 (accessed on 8 November 2017).

17. Sommer, L. Will California Drought Force Changes in Historic Water Rights? *KQED Science.* 11 May 2015. Available online: https://ww2.kqed.org/science/2015/05/11/how-californias-water-rights-make-it-tough-to-manage-drought/ (accessed on 1 August 2017).

18. Wines, M. West's Drought and Growth Intensify Conflict over Water Rights. *The New York Times.* 16 March 2014. Available online: http://www.nytimes.com/2014/03/17/us/wests-drought-and-growth-intensify-conflict-over-water-rights.html (accessed on 30 May 2014).

19. Seager, R.; Hoerling, M. Atmosphere and Ocean Origins of North American Droughts. *J. Clim.* **2014**, *27*, 4581–4606. [CrossRef]

20. Knowles, N.; Cayan, D.R. Potential effects of global warming on the Sacramento/San Joaquin watershed and the San Francisco estuary. *Geophys. Res. Lett.* **2002**, *29*. [CrossRef]

21. Diffenbaugh, N.S.; Swain, D.L.; Touma, D. Anthropogenic warming has increased drought risk in California. *Proc. Natl. Acad. Sci. USA* **2015**, *112*, 3931–3936. [CrossRef] [PubMed]

22. Cooley, H.; Donnelly, K.; Soqo, S.; Bailey, C. *Drought and Equity in the San Francisco Bay Area*; Pacific Institute: Oakland, CA, USA, 2016; p. 24.

23. Feinstein, L.; Phurisamban, R.; Ford, A.; Tyler, C.; Crawford, A. *Drought and Equity in California*; Pacific Institute: Oakland, CA, USA, 2017; p. 71.

24. Reisner, M. *Cadillac desert: The American West and its Disappearing Water*; Penguin Group USA: New York, NY, USA, 1993; ISBN 0-14-017824-4.

25. Brown, F.L.; Ingram, H.M. *Water and Poverty in the Southwest*; University Arizona Press: Tucson, AZ, USA, 1987; p. 226.

26. Ingram, H.M.; Whiteley, J.M.; Perry, R.W. The Importance of Equity and the Limits of Efficiency in Water Resources. In *Water, Place, and Equity*; Whiteley, J.M., Ingram, H.M., Perry, R.W., Eds.; MIT Press: Cambridge, MA, USA, 2008; pp. 1–32. ISBN 978-0-262-23271-5.

27. Wescoat, J.L., Jr.; Halvorson, S.; Headington, L.; Replogle, J. Water, poverty, equity, and justice in Colorado: A pragmatic approach. In *Justice and Natural Resources: Concepts, Strategies, and Applications*; Mutz, K.M., Bryner, G.C., Kenney, D.S., Eds.; Island Press: Washington, DC, USA, 2002; pp. 57–86.

28. Sax, J.L.; Thompson, B.H., Jr.; Leshy, J.D.; Abrams, R.H. *Legal Control of Water Resources: Cases and Materials*; West: Eagan, MN, USA, 2006; ISBN 0-314-16314-X.

29. Brown, F.L. Water markets and traditional water values: Merging commodity and community perspectives. *Water Int.* **1997**, *22*, 2–5. [CrossRef]

30. Whiteley, J.M.; Ingram, H.M.; Perry, R.W. *Water, Place, and Equity*; MIT Press: Cambridge, MA, USA, 2008; ISBN 978-0-262-23271-5.

31. Dill, B.; Crow, B. The colonial roots of inequality: Access to water in urban East Africa. *Water Int.* **2014**, *39*, 187–200. [CrossRef]

32. Goff, M.; Crow, B. What is water equity? *The unfortunate consequences of a global focus on "drinking water."* *Water Int.* **2014**, *39*, 159–171. [CrossRef]

33. McMillan, R.; Spronk, S.; Caswell, C. Popular participation, equity, and co-production of water and sanitation services in Caracas, Venezuela. *Water Int.* **2014**, *39*, 201–215. [CrossRef]

34. Srinivasan, V.; Kulkarni, S. Examining the emerging role of groundwater in water inequity in India. *Water Int.* **2014**, *39*, 172–186. [CrossRef]

35. Tucker, C.M. Creating equitable water institutions on disputed land: A Honduran case study. *Water Int.* **2014**, *39*, 216–232. [CrossRef]

36. Wilder, M.; Liverman, D.; Bellante, L.; Osborne, T. Southwest climate gap: Poverty and environmental justice in the US Southwest. *Local Environ.* **2016**, *21*, 1332–1353. [CrossRef]

37. Syme, G.J.; Nancarrow, B.E.; McCreddin, J.A. Defining the components of fairness in the allocation of water to environmental and human uses. *J. Environ. Manag.* **1999**, *57*, 51–70. [CrossRef]

38. Dunning, H.C. State Equitable Apportionment of Western Water Resources. *Neb. L. Rev.* **1987**, *66*, 76–119.

39. Attfield, R. Environmental Ethics and Intergenerational Equity. *Inquiry* **1998**, *41*, 207–222. [CrossRef]

40. Ingram, H.; Scaff, L.; Silko, L. Replacing Confusion with Equity: Alternatives for Water Policy in the Colorado River Basin. In *New Courses for the Colorado River: Major Issues for the Next Century*; Weatherford, G.D., Brown, F.L., Eds.; University of New Mexico Press: Albuquerque, NM, USA, 1986; pp. 177–200; ISBN 978-0-8263-0854-2.

41. Gleeson, T.; VanderSteen, J.; Sophocleous, M.A.; Taniguchi, M.; Alley, W.M.; Allen, D.M.; YangXiao, Z. Groundwater sustainability strategies. *Groundw. Sustain. Strateg.* **2010**, *3*, 378–379. [CrossRef]

42. Gleeson, T.; Alley, W.M.; Allen, D.M.; Sophocleous, M.A.; Zhou, Y.; Taniguchi, M.; VanderSteen, J. Towards Sustainable Groundwater Use: Setting Long-Term Goals, Backcasting, and Managing Adaptively. *Ground Water* **2012**, *50*, 19–26. [CrossRef] [PubMed]

43. Collins, L. Environmental Rights for the Future? Intergenerational Equity in the EU. *Rev. Eur. Commun. Int. Environ. Law* **2007**, *16*, 321–331. [CrossRef]

44. Peterson, J.M.; Marsh, T.L.; Williams, J.R. Conserving the Ogallala Aquifer: Efficiency, Equity, and Moral Motives. *Choices* **2003**, *18*, 15–18.

45. Weiss, E.B. In Fairness to Future Generations. *Environ. Sci. Policy Sustain. Dev.* **1990**, *32*, 6–31. [CrossRef]

46. Collins, L.M. Revisiting the Doctrine of Intergenerational Equity in Global Environmental Governance. *Dalhous. Law J.* **2007**, *30*, 79–134.

47. Stroshane, T. *Drought, Water Law, and the Origins of California's Central Valley Project*; University of Nevada Press: Reno, NV, USA, 2016; ISBN 978-0-87417-001-6.

48. U.S. Bureau of Reclamation. *Second Amended Contract for Exchange of Waters*; U.S. Bureau of Reclamation: Washington, DC, USA, 1967.

49. U.S. Bureau of Reclamation. Sacramento River Settlement Contractors Contract Renewal Effort. Available online: https://www.usbr.gov/mp/cvpia/3404c/srsc/index.html (accessed on 19 November 2017).

50. U.S Bureau of Reclamation. *Contract between the United States and Glenn-Colusa Irrigation District, Diverter of Water from Sacramento River Sources, Settling Water Rights Disputes and Providing for Project Water*; U.S. Department of the Interior: Sacramento, CA, USA, 2005.

51. U.S. Bureau of Reclamation. *Central Valley Project Municipal and Industrial Water Shortage Policy Guidelines and Procedures*; U.S. Department of the Interior: Sacramento, CA, USA, 2017.

52. U.S. Bureau of Reclamation. *Long-Term Renewal Contract Between the United States and Tulare Irrigation District Providing For Project Water Service From Friant Division*; Central Valley Project: Sacramento, CA, USA, 2001; p. 62.

53. U.S. Bureau of Reclamation. *Central Valley Project (CVP) Water Supply for 2016*; U.S. Department of the Interior: Sacramento, CA, USA, 2016.

54. U.S. Bureau of Reclamation. *California Department of Water Resources Central Valley Project and State Water Project 2016 Drought Contingency Plan for Water Project Operations February–November 2016*; U.S. Department of the Interior: Sacramento, CA, USA, 2016.

55. State Water Resources Control Board. *2015 Summary of Water Shortage Notices*; State Water Resources Control Board: Sacramento, CA, USA, 2015.

56. Boxall, B. California moves to restrict water pumping by pre-1914 rights holders. *Los Angeles Times.* 12 June 2015. Available online: http://www.latimes.com/local/lanow/la-me-ln-drought-water-rights-20150612-story.html (accessed on 19 November 2017).

57. Joyce, E. No Fed Project Water for Some California Farmers. Available online: http://www.capradio.org/70179 (accessed on 4 October 2017).

58. Cooley, H.; Donnelly, K.; Phurisamban, R.; Subramanian, M. *Impacts of California's Ongoing Drought: Agriculture*; Pacific Institute: Oakland, CA, USA, 2015; p. 24.

59. Howitt, R.E.; Medellin-Azuara, J.; MacEwan, D.; Lund, J.R.; Sumner, D.A. *Economic Analysis of the 2014 Drought for California Agriculture*; Center for Watershed Sciences: Davis, CA, USA, 2014; p. 20.

60. California Budget Project. Poverty varies significantly across California counties—Percentage of people with incomes below the federal poverty line, 2008–2012. 2014. Available online: http://calbudgetcenter.org/wp-content/uploads/140108_poverty_table_and_map.pdf (accessed on 17 November 2017).

61. Choy, J.; McGhee, G. Groundwater: Ignore It, and It Might Go Away. *Understanding California's Groundwater.* 31 July 2014. Available online: http://waterinthewest.stanford.edu/groundwater/overview/ (accessed on 21 November 2017).

62. Sabalow, R.; Kasler, D.; Reese, P. Farmers say, "No apologies," as well drilling hits record levels in San Joaquin Valley. *The Sacramento Bee.* 25 September 2016. Available online: http://www.sacbee.com/news/state/california/water-and-drought/article103987631.html (accessed on 18 November 2017).

63. Siegler, K. California's War over Water Has Farmer Fighting Farmer. *All Things Considered.* 4 June 2015. Available online: https://www.npr.org/sections/thesalt/2015/06/04/411475620/californias-war-over-water-has-farmer-fighting-farmer (accessed 18 November 2017).

64. U.S. Bureau of Reclamation News Archive: Reclamation Makes Historic Releases of Water from Friant Dam to the San Joaquin River Exchange Contractors Due to Drought. Available online: https://www.usbr.gov/newsroom/newsrelease/detail.cfm?RecordID=46887 (accessed on 9 November 2017).

65. Grossi, M. Federal court rejects bid to stop flow from Friant Dam. *Fresno Bee.* 27 May 2014. Available online: http://www.fresnobee.com/news/local/water-and-drought/article19520583.html (accessed on 19 November 2017).

66. Western Farm Press Staff. Friant Water Authority Challenges USBR Decision in Court. *Western Farm Press.* 20 May 2014. Available online: http://www.westernfarmpress.com/irrigation/friant-users-challenge-usbr-decision-court (accessed on 19 November 2017).

67. California Water Districts Voluntarily Drop Lawsuit against Bureau of Reclamation. *Lexis Legal News.* 17 June 2015. Available online: https://www.lexislegalnews.com/articles/591/california-water-districts-voluntarily-drop-lawsuit-against-bureau-of-reclamation (accessed on 18 November 2017).

68. Griswold, L. San Joaquin Valley agencies sue feds for $350 million over no water deliveries. *Fresno Bee.* 5 October 2016. Available online: http://www.fresnobee.com/news/local/article106312712.html (accessed on 28 June 2017).

69. United States Sued for $350M in Damages from 2014 Water Diversion. *Lexis Legal News*, 6 October 2016.

70. Rauch Communication Consultants Inc. *Draft 2016 Strategic Plan Update & 2017 Implementation of the Plan*; Friant Water Authority: Lindsay, CA, USA, 2017; p. 13. Available online: https://static1.squarespace.com/static/58c2eccc15d5db46200ea426/t/59304db03e00be246227b128/1496337844929/FWA_2016-Strategic-Plan-Draft_201612.pdf (accessed on 8 November 2017).

71. Wiel, S.C. Theories of Water Law. *Harv. Law Rev.* **1914**, *27*, 530–544. [CrossRef]

72. Pitzer, G. Drought Puts California Water Rights in Crosshairs for Reform. *Water Deeply.* 1 September 2015. Available online: https://www.newsdeeply.com/water/articles/2015/09/01/drought-puts-california-water-rights-in-crosshairs-for-reform (accessed 22 October 2017).

73. Pisani, D.J. Enterprise and Equity: A Critique of Western Water Law in the Nineteenth Century. *West. Hist. Q.* **1987**, *18*, 15–37. [CrossRef]

74. Hundley, N. *The Great Thirst: Californians and Water—A History*; Revised Edition; University of California Press: Berkeley, CA, USA, 2001; ISBN 978-0-520-22455-1.

75. Wilkinson, C.F. Aldo Leopold and Western Water Law: Thinking Perpendicular to the Prior Appropriation Doctrine. *Land Water Law Rev.* **1989**, *24*, 1–38.

76. Sugg, Z.P.; Ziaja, S.; Schlager, E.C. Conjunctive groundwater management as a response to socio-ecological disturbances: A comparison of 4 western US states. *Tex. Water J.* **2016**, *7*, 1–24.

77. Scanlon, B.R.; Faunt, C.C.; Longuevergne, L.; Reedy, R.C.; Alley, W.M.; McGuire, V.L.; McMahon, P.B. Groundwater depletion and sustainability of irrigation in the US High Plains and Central Valley. *Proc. Natl. Acad. Sci. USA* **2012**, *109*, 9320–9325. [CrossRef] [PubMed]

78. Smith, R.G.; Knight, R.; Chen, J.; Reeves, J.A.; Zebker, H.A.; Farr, T.; Liu, Z. Estimating the permanent loss of groundwater storage in the southern San Joaquin Valley, California. *Water Resour. Res.* **2017**, *53*, 2133–2148. [CrossRef]

79. Than, K. Groundwater overuse cuts San Joaquin Valley's water storage ability. *Stanford News.* 12 April 2017. Available online: https://news.stanford.edu/2017/04/12/groundwater-overuse-cuts-san-joaquin-valleys-water-storage-ability/ (accessed on 4 December 2017).

80. Tarlock, A.D. The law of equitable apportionment revisited, updated, and restated. *U. Colo. L. Rev.* **1985**, *56*, 381–410.

81. *Nebraska v. Wyoming, 325 U.S. 589*; U.S. Supreme Court: Washington, DC, USA, 1945.

82. Ruhl, J.B. Equitable Apportionment of Ecosystem Services: New Water Law for a New Water Age. *J. Land Use Environ. Law* **2003**, *19*, 47–57.

83. Gray, B.; Hanak, E.; Frank, R.; Howitt, R.; Lund, J.; Szeptycki, L.; Barton, T. *Allocating California's Water: Directions for Reform*; Public Policy Institute of California: San Francisco, CA, USA, 2015.

84. Northern California Water Association. *Drought Planning in the Sacramento Valley: Recommendations for 2015*; Northern California Water Association: Sacramento, CA, USA, 2014.

85. Miller, J.M. When Equity is Unfair-Upholding Long-Standing Principles of California Water Law in City of Barstow v. Mojave Water Agency Casenote. *McGeorge Law Rev.* **2000**, *32*, 991–1018.

86. *National Audubon Society v. Superior Court (1983) 33 Cal.3d, 419, 189 Cal.Rptr. 346; 658 P.2d 709*; Supreme Court of California: San Francisco, CA, USA, 1983.

87. Gray, B.E. In Search of Bigfoot: The Common Law Origins of Article X, Section 2 of the California Constitution Water. *Hastings Const. Law Q.* **1989**, *17*, 225–274.

88. California Constitution. Article X Water § 2. 1976. Available online: https://leginfo.legislature.ca.gov/faces/codes_displayText.xhtml?lawCode=CONS&division=&title=&part=&chapter=&article=X (accessed on 24 January 2018).

89. Arnold, C.A. Adaptive Water Law. *Univ. Kans. Law Rev.* **2013**, *62*, 1043–1090.

90. Hanemann, M.; Dyckman, C. The San Francisco Bay-Delta: A failure of decision-making capacity. *Environ. Sci. Policy* **2009**, *12*, 710–725. [CrossRef]

91. Shilling, F.M.; London, J.K.; Liévanos, R.S. Marginalization by collaboration: Environmental justice as a third party in and beyond CALFED. *Environ. Sci. Policy* **2009**, *12*, 694–709. [CrossRef]

92. Hanemann, M.; Dyckman, C.; Park, D. California's Flawed Surface Rights. In *Sustainable Water: Challenges and Solutions from California*; Lassiter, A., Ed.; University of California Press: Berkeley, CA, USA, 2015; pp. 52–83. ISBN 978-0-520-96087-9.

93. Kallis, G.; Kiparsky, M.; Norgaard, R. Collaborative governance and adaptive management: Lessons from California's CALFED Water Program. *Environ. Sci. Policy* **2009**, *12*, 631–643. [CrossRef]

94. Lejano, R.P.; Ingram, H. Collaborative networks and new ways of knowing. *Environ. Sci. Policy* **2009**, *12*. [CrossRef]

95. Gulley, R.L. *Heads above Water*; Texas A&M University Press: College Station, TX, USA, 2015; ISBN 978-1-62349-269-4.

resources

MDPI

Article

The Vital Minimum Amount of Drinking Water Required in Ecuador

Andrés Martínez Moscoso [1,2,*], Víctor Gerardo Aguilar Feijó [1] and Teodoro Verdugo Silva [1,2]

1   Faculty of Economics and Administration, University of Cuenca, Cuenca 010203, Ecuador;
    victor.aguilar@ucuenca.edu.ec (V.G.A.F.); teodoro.verdugo@ucuenca.edu.ec (T.V.S.)
2   Faculty of Jurisprudence, Political and Social Sciences, University of Cuenca, Cuenca 010203, Ecuador
*   Correspondence: andres.martinez@ucuenca.edu.ec; Tel.: +593-7405-100 (ext. 2280)

Received: 2 November 2017; Accepted: 13 February 2018; Published: 24 February 2018

Abstract: In 2017, the government of Ecuador established the minimum quantity of water required to be provided for free by drinking water utilities. Ecuador recognized the access to water as a fundamental human right because it guarantees the good living, known as "Sumak kawsay", an indigenous Andean concept, in the Ecuadorian Constitution. This represents a novel approach to water rights in the world, as it is the first attempt to establish a minimum quantity of water under a constitutional guarantee by legislation, rather than regulation or judicial decision. However, this novel legislative approach raises the question of how this minimum amount of free water will impact the most vulnerable members of the Ecuadorian community. This paper provides the results of the first comprehensive research of the minimum required water provision in Ecuador. In order to measure the impact on the income of households, we built a methodology integrating: doctrinaire analyses, normative studies, and economic analyses. According to the Ecuadorian legislation, over-consumption of raw water generates additional costs that must be paid by water companies to the central government. In that regard, there is an inevitable relationship between the efficiency of the service and those additional costs. Efficiency, on this case, is the capacity of water companies (public or private) to provide water services at an adequate price, observing the following parameters: quantity, quality and sufficiency. Our research found that with this legislation in three Ecuadorian local governments (Cuenca, Gualaceo and Suscal), the most vulnerable households (i.e., low-income and/or indigenous households) will be affected the most. This means that and those families will spend the most part of their income on water services otherwise they would have to reduce their water consumption.

Keywords: drinking water; minimum vital of drinking water; human right to water; Ecuador; family income; public services

---

1. Introduction

From a normative standpoint, the case of Ecuador presents a peculiarity in relation to water management. Until 1960, water administration remained private; the State simply intervened to resolve disputes and authorize usage. That year water was declared a national good, a principle ensured in 1972 when all continental waters of the country were declared public domain. In 1993, the process of State modernization affected the provision of public services, including water provision and management which was delegated to the National Council of Water Resources (1994) [1]. In 2008 Ecuador's new Constitution reconfirmed the State as primary authority for water management, conservation, recovery, integrated management of water resources, watersheds and ecological flows associated to the water cycle (Art. 411 and 412).

An important feature of the Ecuadorian case is that irrigation water regulations have been prioritized over drinking water regulations due to the influence of 1970s Agrarian Reform, which generated a policy of subsidies and debt write-offs in favor of indigenous and peasant groups.

This has produced an important gap between water access, efficiency and quality in rural and urban areas. With the new constitutional framework and the creation of the National Water Authority (SENAGUA) in 2008, access to drinking water has been increased (from 69% in 2006 to 83.6% in 2016). In the rural areas, however, the access to water through public connection is only 59%.

In May of 2017, SENAGUA signed the Ministerial Agreements No. 2017-1522 and 2017-1523, which approved the formula to calculate the cost of raw water, as well as the minimum amount of drinking water (200 L) guaranteed under the 2008 Ecuadorian Constitution. As positive as the recognition of the right to water is, quantifying a vital minimum amount of water per person poses the following complex questions: first, is the standardization made in Ecuador correct, since it doubles the international average? Second, is the application of the vital minimum of water widening the differences between subnational governments regarding the provision of water, quality of service, financial capacity of the lending companies, and quality of life for families?

The purpose of this research is to determine the extent to which the application of these Ministerial agreements increases (or reduces) the differences between local governments, in terms water provision efficiency, and the economic capacity of the most vulnerable families to pay for water exceeding the Constitutionally-guaranteed vital minimum amount. The study selected three different municipalities in the austral area of Ecuador (See Figure 1): Cuenca the third largest city in the country, with 591.996 inhabitants, recognized for its efficiency in the provision of public services [2,3], Gualaceo and intermediate city with 47.411 inhabitants [3], and Suscal a small city of 5.998 inhabitants, with an indigenous majority (data based on [3,4]). The working hypothesis is that the application of the Minimum Amount of Raw Water (MARW) negatively affects the three subnational governments, both suppliers and most vulnerable consumers. We expect a greater effect in municipalities that are less efficient in the provision of potable water services and that have a socially and economically vulnerable population.

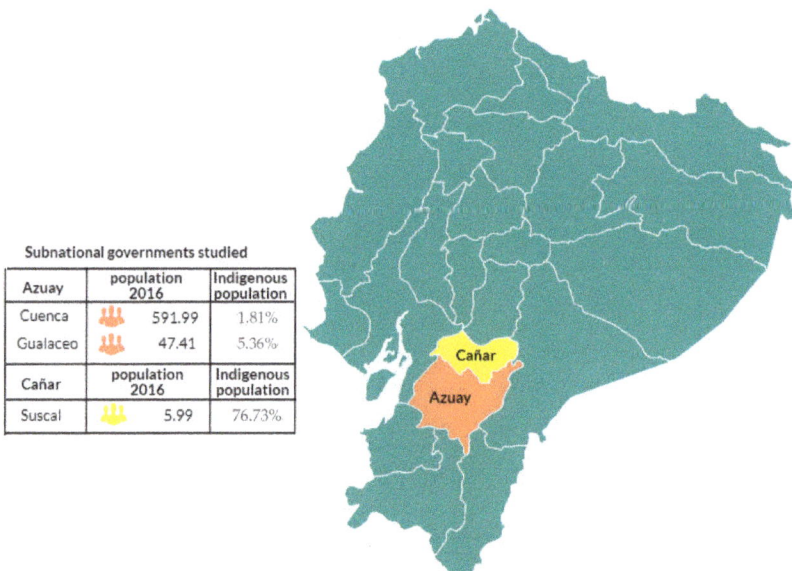

Figure 1. Subnational governments studied and population in thousands of inhabitants. Source: Prepared by the Authors based on population projections of INEC [4].

2. Background

According to the World Health Organization (WHO), access to the vital minimum amount of drinking water is directly related to hygiene and public health. Thus, ensuring a minimum access to water is crucial.

The WHO states that there should be a minimum acceptable quantity of water to cover basic needs, such as drink, food preparation and hygiene. In order to measure water requirements with provision of water services and its impact on healthcare, the WHO establishes four service levels (see Table 1): service without access, basic access, intermediate access and optimal access [5]. The first level, service without access (less than 5 L per day), cannot guarantee minimum hygiene conditions and entails a high risk for poor health. Basic access (20 L per day) guarantees basic hygiene (hand-washing), although the risk for poor health remains high. Intermediate access (50 L per day) assures basic personal and food hygiene and health. Finally, optimal access (100 L per day and more), satisfies all hygiene needs.

Table 1. The vital minimum of drinking water (World Health Organization).

Service Level	Access Level	Level of Effect upon Health
Without access	Less than 5 L per day	Cannot guarantee minimum hygiene conditions and entails a high risk for poor health.
Basic access	20 L per day	Guarantees basic hygiene (hand-washing), although the risk for poor health remains high.
Intermediate access	50 L per day	Assures basic personal and food hygiene and health risks are low.
Optimal access	100 L per day or more	Satisfies all hygiene needs, with a very low risk to health.

Source: Prepared by the Authors based on [5].

Access to water for domestic consumption is defined by WHO as: "water used for all usual domestic purposes including consumption, bathing and food preparation", taking into account the following categories: drinking water (for drinking and cooking); water for personal hygiene (basic needs and personal care); and, water for domestic cleaning (car-washing, garden-watering, etc.) [5]. However, authors such as Thompson [6], suggest that a fourth category, "productive use", should be included. This category should be applied to households with depressed economies in developing countries, in reference to the water that is used for small-scale agriculture and livestock, construction, etc. (Falkenmark argues that 1369 L/c/d (500 m^3/c/y) is the minimum required to run a modern society living in semi-arid conditions, with 1095 L/y) required for irrigation, and 274 L/c/d (100 m^3/c/y) for domestic and industrial needs).

The United Nations Children's Fund (UNICEF) and World Health Organization (WHO), through their Joint Monitoring Program, define reasonable access to water as "the availability of at least 20 L per person per day from a source within one kilometer of the consumer's home" [6]. However, this definition refers to access, not to the quantity that is recommended in liters per capita per day (LPCPD). The discussion regarding the minimum amount of water needed by an individual includes other variables such as the type of activity that is performed, the temperature, the geographical conditions, etc. [7].

Regarding the protection of rights related to water resources, the 108th plenary meeting of the General Assembly of United Nations (With 122 votes in favour, no votes against, and 41 abstentions), held in 28 July 2010, recognized drinking water and sanitation as human rights. In this sense, Ecuador was ahead of the region by innovatively recognizing water as a fundamental and inalienable human right in the 12th article of the 2008 Constitution. Water is also considered a strategic national heritage of public use, and above all, essential for life. Other countries that made the same recognition prior to 2010 are South Africa (1997) ("Other countries have recognised the right to a healthy environment in their constitutional texts. Whilst over 60 constitutions refer to environmental obligations, less than a half expressly refer to the right of the citizens to a healthy environment. Only the South African Bill of Rights enshrines an explicit right to access to sufficient water" [8]), Uruguay (2004), and Bolivia (2009) [9].

When applying human rights principles to water and sanitation, the following concepts should be observed: non-discrimination and equality; access to information and transparency; participation; accountability, and sustainability. It also should include: availability, physical accessibility, quality and safety, affordability, acceptability, dignity, and privacy [10,11].

The Ecuadorian Constitution defines water as: (a) a human right; (b) a national heritage and strategic sector; (c) a public service; and (d) an element to achieve food and energy sovereignty. As a human right, it is the State's primary duty to guarantee water (Art. 3, No. 1.) In fact, water is regarded as a fundamental and inalienable human right protected under the right to well-being (Art. 12). This implies that it cannot and should not be isolated but understood in interrelation with other rights, by virtue of its interdependent and indivisible character (article 11, No. 6), especially in connection to the right to health (article 32) and to a dignified life (Art. 66).

As a strategic national patrimony of public use, water is inalienable, inseparable, non-excludable, and essential for life (Art. 12). Decision and control is the exclusive responsibility of the State, which administers, regulates, controls, and manages water in accordance with the principles of environmental sustainability, precaution, prevention, and efficiency (Art. 313). The Constitution states that water is vital for nature and for the existence of human beings and, therefore, cannot be privatized (Art. 318).

As a resource to achieve energy sovereignty, constitutional provisions establish that this sovereignty shall not be reached in detriment of, or by compromising, water supplies. To this effect, the State will promote, in the public and private sectors, the use of environmentally clean technologies, and non-polluting and low-impact alternative energies (Art. 15). Similarly, water is a development goal in Ecuador. The Constitution states that one of the goals is to recover and conserve nature in order to maintain a healthy and sustainable environment, as means to provide people and communities an equitable, permanent, and high quality access to water, air, and soil (article 276, No. 4).

Water and its preservation are also the responsibility of the State in order to achieve food sovereignty. To ensure this, the State must promote redistributive policies, which give peasants access to land, water, and other productive resources (Article 281 No. 4), bearing in mind that it is the State's obligation to avoid the monopolization or privatization of water and its sources. In addition, the State shall regulate the use and management of irrigation water for food production, under the principles of equity, efficiency, and environmental sustainability (Article 282).

Finally, the Ecuadorian Constitution considers water and its provision a public service (Art. 314.) In fact, water sanitation, supply, and irrigation can only be provided by national or communal entities. The State will encourage alliances between public and community sectors to provide water related services (Art. 318). By means of a constitutional clause, municipal governments of Ecuador are assigned the responsibility of providing potable water, sewage and wastewater treatment, and are forbidden to suspend any of these services (Art. 264 No. 4; Art. 326 No. 15.) It is the obligation of the State, at all the levels of government, to guarantee the uninterrupted provision of public drinking water services (Art. 375 No.6).

3. State of the Art

On the 28 July 2010, the United Nations General Assembly recognized the right to water, sanitation and hygiene (WASH). WASH was granted the status of an essential human right, since it guarantees both the enjoyment of life, as well as other human rights. In its resolution 64/292, the Assembly called upon the subjects of international law to provide resources, and technical and technological assistance, especially for developing countries, so that their populations can access drinking water and sanitation. Such access must be sustainable, since there is no point in having access to water through platforms and infrastructure if this is not sustainable over time, due to low quality or high maintenance costs [10]. If this condition is not met individuals will become unprotected again and, consequently, will lack access to water.

The Inter-American Human Rights System, through the Inter-American Court, Inter-American Commission, and its case law, has developed two transcendental criteria in the field of drinking water. The first deals with quantity and quality of water for the development of a dignified life, in the case of the Xákmok Kásek Indigenous Community vs. Paraguay, of 24 August 2010. The second relates to the minimum living conditions for persons deprived of their liberty, in in the cases Vélez Loor vs. Panama, date 23 November 2010, and Pacheco Teruel et al. vs. Honduras, of 27 April 2012. This last criterion establishes that the State must provide drinking water to all inmates to guarantee a dignified life.

In general, the application of the human right to drinking water and sanitation is progressive. Each State must develop and expand its coverage so that all inhabitants can gradually have access to these services, avoiding a setback for human rights in this area. However, as pointed out [10] in the Special Rapporteur's report on these rights: the main risks, especially in times of crisis, are the budget cuts that affect the development of infrastructure and provision of public services. This directly impacts human rights, such as the right to health, since people with fewer resources ("The cuts in public spending particularly affect the poorest and most marginalized, whose income usually comes mainly from social benefits, which depend heavily on public services and devote a greater part of their income to basic services" [10]) are unable to access quality water or have their service interrupted due to lack of payment. Given disproportionate increases in rates, the lack of income from unemployment during such times leaves vulnerable groups without the resources to pay for essential services.

In 2007, Ecuador presented the "II National Report on the Millennium Development Goals" (The report was prepared by the Millennium Social Research Center (CISMIL), composed of the United Nations Development Program (UNDP), the Latin American Faculty of Social Sciences, FLACSO in Ecuador, and the National Secretary for Planning and Development (SENPLADES)). In it, the National Secretary of Planning and Development acknowledged that while progress had been made in access to drinking water and sanitation, the territorial gaps (urban/rural, coast, highlands and Amazon Region) are still considerable. Figure 2 shows the evolution of drinking water access in Ecuador since 1995 when only 37% of Ecuadorian households had access to water through the public network. This percentage grew to 67% in 2001 and to 71.98% 2010. Ecuador's drinking water coverage has improved in the last decade, from 69% (2006) to 83.6% (2016), thanks to the significant investment in accessibility by the Local Governments, which have the capability to provide the service. According to the statistics from the Joint Monitoring Program (JMP) of WHO and UNICEF in 2015, coverage levels of drinking water services in Ecuador have increased to 92% at a national level, 97% in urban areas, and 82% in the rural areas.

3.1. Analysis of International and National Case Law on the Vital Minimum Amount of Drinking Water

Regarding international case law, there are two cases concerning the determination of a vital minimum of drinking water. The first, in South Africa, of Lindiwe Mazibuko and others against the city of Johannesburg, in 2009, in which two issues were discussed: (a) if the basic free water supply, that is, 6 kiloliters per month for each connection holder (or, 60 L per person per day), was in conflict with the Constitution; and, (b) if the installation of prepaid water meters was legal, as they worked with coins resembling parking meters, for the excess rate. The Constitutional Court considered that as there were too many answers to the question of what constitutes a sufficient amount of water, therefore, a specific vital minimum should not be determined, since that decision corresponded to the legislator or the government as a matter of public policy.

The second case corresponds to the Constitutional Court of Colombia, in its Judgment T-740, of 2011. The Sentence determined that the vital minimum of drinking water is 50 L per person per day. The constitutional judge developed the fundamental right to water applying the Resolution of the General Assembly of the United Nations as "soft law" for the legal system, obliging the Operator to reconnect, restore the service, and review payment agreements so that the user could afford them. It also mandated the installation of a flow redactor which guarantees the minimum water for consumption (50 L). Finally, it ordered the national government, through the local government, to subsidize 50% of the costs, for the user to pay for the service.

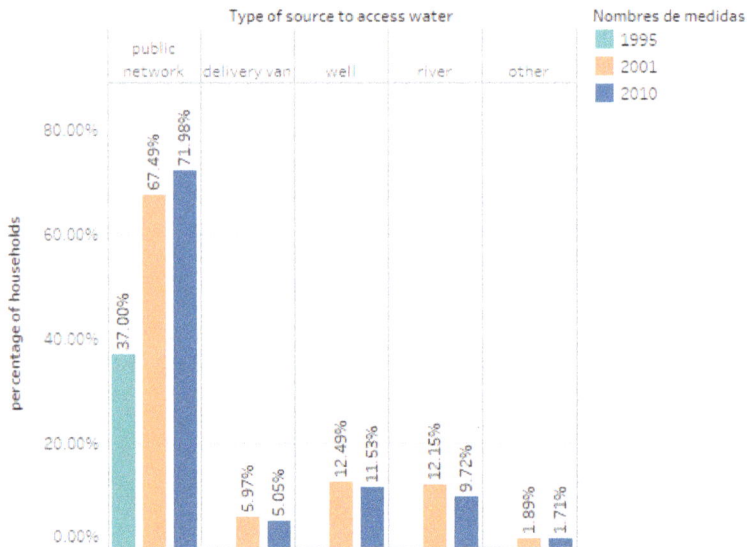

Figure 2. Access to water in Ecuadorian households by type of source (1995–2010). Source: Prepared by the Authors based on [12,13].

The Colombian decision is based on the following criterion: "… The deprivation of drinking water entails a serious violation of the state obligations derived from the fundamental right to water, specifically those of availability and accessibility, inasmuch as, first, it restricts the possibility that this sector of the population, which is in a circumstance of manifest weakness, has access to the services and facilities of water, and second, the availability of water for the satisfaction of personal and domestic needs, such as the preparation of food, personal hygiene and household hygiene. Thus, in the event of failure to pay more than two consecutive billing periods, the public water utility should, as has been pointed out in this Corporation's case law, report the credit situation of the user, and the procedure to follow, so that she can keep up to date with her obligations …" (Full sentence available in [14]).

In the Ecuadorian case, the Constitutional Court ruled on the management and administration of strategic sectors, through decision No. 001-12-SIC-CC. This rule absolves the President's consultation on the content of the articles 313, 315 and 316 of the Constitution. The consultant, after stating that during the 1990s there was a privatizing tendency from the State, manifests that the current Constitution seeks to guarantee the satisfaction of collective needs through the priority intervention of the State in the management of strategic sectors, as well as public services. Distinguishing between the administration, regulation, and control of those sectors, which corresponds to the Central State according to Art. 313 of the Constitution; and the management which corresponds to public companies, as established in Art. 315 of the same legal body [15].

For this reason, two doubts arise regarding the application of the mentioned articles. The first is that public companies, being part of the State, either enjoy or lack administrative, regulatory, and control powers in strategic sectors and public services. The second refers if the rest of the state organizations (such as any Ministry), have to create a public company, if they want to manage a strategic sector.

Regarding the first consultation, the Constitutional Court mentions Art. 313 of the Constitution, which establishes the exclusive power of the State to manage, regulate, control and operate the strategic sectors. Although these powers have been granted to the State, it is possible to distinguish between the central State and public companies. The first is responsible for the administration, regulation, and control through the competent public administration authorities, in accordance with the law. In addition, the seconds, the public companies which are authorized only for the management of a strategic sector, and in every case it is obligatory to obtain the authorization from the central State (Regulatory and control organs) [15].

Article 316 establishes a priority order for the management of strategic sectors and the provision of public services: State; joint ventures in which the state has a majority shareholding; private initiative and popular and solidarity economy. We have, therefore, the State's own management, and delegation, which can appear in two cases. The first refers to the aforementioned joint ventures, and the second, to private entities [15]. The latter is exceptional and only applies when it is necessary to satisfy the general interest, and the State does not have sufficient technical or economic capacity, or when demand cannot be met by public or mixed companies [16].

Therefore, the Court concludes that the power attributed to the State in Art. 313 manifests itself through two concrete functions. The first is the authorization granted by the central State to public companies, to manage strategic sectors and provide public services. The second is in the delegation that the State can grant to non-public companies to perform such management [15].

Regarding the second consultation, the Court states that when State institutions, as a means to provide public services inherent to their competence, need to manage strategic sectors, they will not need to establish public companies for that purpose. It will only be necessary to have access to the corresponding enabling titles granted by the regulatory and control authorities [15]. Thus, the Ministries, not being public companies, can request such titles themselves.

3.2. The Vital Minimum Quantity of Drinking Water in the Ecuadorian Legislation

The Organic Law on Water Resources, Uses and Exploitation of Water, in defining the human right to water, refers to its availability. According to the Law water should be clean, sufficient, healthy, acceptable, accessible and affordable, for personal and domestic use. Then the vital or "vital minimum" amount of water is important since no person can be deprived, excluded or stripped of this right. It can be considered as the cornerstone on which the enforceability of the human right to water is based. In this way, raw water destined to be processed within this range is free.

Article 18 defines the competence and attribution of the Single Water Authority to set the rates to authorize the use of water. In this context, one day before the end of the last Presidential term (23 May 2017), the Secretary of Water signed Ministerial Agreements No. 2017-1522 and 2017-1523. These agreements established the calculation formula to obtain the raw water referential rate, fixing the minimum vital amount of water in 200 L per capita per day as a minimum amount unified value). Later, the new Administration (2017–2021) amended Ministerial Agreement No. 2017-1522 through Agreement No. 2017-0010, dated 28 June 2017. It established a new administration for SENAGUA—now presided by a former indigenous leader, linked to community water management, which specifically reformed the rates to authorize the use of water.

The raw water referential rate is based on the formula of total costs of water sustainability, over the authorized national volume of water, which has duration of 5 years. However, the most important thing is the differentiation made in the rate according to the type of use and/or utilization, where those who use less than 5 L per second for activities linked to food sovereignty are exempt from payment; while rates are significantly taxing for productive irrigation, industrial use, tourism, hydroelectricity and water bottling.

As for Agreement No. 2017-1523, it establishes the minimum vital amount of water in Ecuador at an equivalent of 200 L of raw water per inhabitant per day. Therefore, the excess consumed above the vital minimum should be charged for. The charging entities are the providers of drinking water services at a national level (Municipal Government or its Public Companies, and Drinking Water Boards).

4. Methodology

We developed our investigation according to the Ecuadorian Legal Framework. The article 59 of the Organic Law on Water Resources, Uses and Exploitation of Water establishes that:

"The vital quantity of raw water destined for processing for human consumption is free to guarantee the human right to water. When the established minimum vital amount is exceeded, the corresponding rate will be applied."

In this context, the agreement No. 2017-1523, issued by the National Water Authority, in its article 3 indicates that: operators who exceed the minimum vital amount of raw water to supply households, incur a cost of USD $0.0039 per m^3.

We developed the following variables and indicators to determine the extent to which the application of the minimum vital amount of drinking water established by the government affects local governments, in terms of water provision efficiency, and the economic capacity of the most vulnerable families.

4.1. Determination of the Minimum Vital Consumption

The value representing the minimum vital consumption follows the next scheme:

Consumption is distributed in categories which are called activities. According Table 2, the total of activities amounts to 154 LPCPD, a quantity to which an additional amount of 46.2 LPCPD is added, which represents a 30% of that total, up to 200 LPCPD. This 30% does not constitute a consumption activity; it directly reflects the raw water linked to the operation of the system. In this sense, an operator depending on its level of efficiency, exceed, match, or reduce that amount. When the amount is exceeded, the excess over 200 LPCPD generates a cost of USD $0.0039 per cubic meter which must be reverted to the State, according to the regulations, or charged to the household.

Table 2. Minimum vital amount of raw water. Based on [17].

Activity	Consumption (L/h/d)	Justification
Larger body hygiene (shower)	80	2 × 8 L/m × 5 min (expenditure 0.134 L/s)
Minor body hygiene (washing hands, teeth, etc.)	9	average referential value
Toilet	24	4 flushes × 6 L
Laundry	20	average referential value
Dish washing	9	average referential value
Consumption, cooking	12	average referential value
Sub Total	154	
Increase due to operation of the system, trade, industries and losses (30%)	46.2	
Total	200.2	
Vital Amount of Adopted Raw Water	200	

4.1.1. MARW's Impact on Service Providers

In order to measure the impact of Minimum Amount of Raw Water (MARW) on the providers of drinking water, Formulas (1) and (2) apply to each local governments.

$$RWE\ (m^3) = RW - MARW \tag{1}$$

where Raw water excess RWE (m^3): raw water excess in m^3, measures the difference between the m^3 of raw water that a local governments needs in order to satisfy the average present or typical drinking water consumption of a household, and the minimum vital amount of raw water consumption in m^3

according to the regulations of the Secretary of Water. RW: raw water, refers to the quantity in m³ of raw water that a local governments needs in order to satisfy the average current or typical drinking water consumption monthly of a household. MARW: is the minimum vital amount of raw water consumption monthly in m³ which a household needs on average.

$$RWE\ (USD) = RWE\ (m^3) \times 0.0039USD \tag{2}$$

where RWE (USD): raw water excess in USD $, measures the cost of excessive consumption of raw water in USD $ per household to be reverted to the State. This results from the product between the excess consumption of raw water in m³ times the amount of m³ which exceeds the MARW of 200 LPCPD (USD $0.0039).

4.1.2. MARW's Impact on Households Living in Extreme Poverty:

As mentioned, the excess, once monetized, must be reverted to the State. The operator of the service in each local governments can cover such value without using the rate, or, alternatively, add that cost to the value paid by the consumer, thus giving it the appearance of a cost to cover. When municipalities index the costs of excess raw water over the rate to be paid by households, the impact of the vital minimum consumption of raw water on the income of the households living in extreme poverty [6] is determined by applying the Formulas (3)–(5).

$$TAC(\%\ AI) = \frac{(TACPC(USD)}{(AIPC(USD))} \tag{3}$$

where TAC(% AI): Typically Average Consumption of drinking water (TAC) as percentage of average income, measures the monthly consumption of drinking water per capita in monetary units as a percentage of the average per capita monthly income of a household considered to be living in extreme poverty. TACPC(USD): Typically Average Consumption Per Capita (TACPC), refers to the monetary value of the average monthly consumption of drinking water in a household, divided by the average number of people in a household. Average Income Per Capita (AIPC) (USD): is the average monthly income per capita of a household living in extreme poverty.

$$TCPC(\%\ AI) = \frac{(TACPC(USD) + RWEPC(USD))}{(AIPC(USD))} \tag{4}$$

where TCPC(% AI): is the total cost per capita of the average monthly consumption of drinking water as a percentage of average income, and it measures the average monthly per capita consumption of drinking water in the monetary units that a household pays to the potable water service operators, plus the cost of the monthly excess of consumption of raw water in monetary units per capita under the assumption that the household assumes the cost; divided by the average monthly income per capita of a household considered to be poor. TACPC(USD): typically average consumption per capita, refers to the monetary value of the average monthly consumption of drinking water in a household, divided by the average number of people in a household. RWEPC(USD): Raw Water Excess Per Capita (RWEPC) in USD $, measures the cost of excess consumption of raw water in monetary units per capita. It results from the product between the excess consumption of raw water in m³ times the value of m³ which exceeds the MARW of 200 LPCPD (USD $0.0039), divided by the number of average people per household.

$$IH = \frac{(TCPC(\%\ AI) - TAC(\%\ AI))}{(TAC(\%\ AI))} \tag{5}$$

where IH: impact household, measures the impact on a household living in extreme poverty (According to the National Institute of Statistics and Census of Ecuador, an individual is extremely poor by income, when the total per capita income is below a minimum income to not be considered as extremely poor, that is, when it is below the extreme poverty line, which for December 2016 was $47.72 [18]), as well as the growth rate represented by the additional cost that the household should pay for excess water consumption. TCPC(% AI): as is the total per capita cost of the average monthly consumption of drinking water as a percentage of average income, which includes the cost of excess water consumption. TAC(% AI): typically average consumption of drinking water as percentage of average income.

4.2. Data

The methodology described in the previous section is applied with the information from Table 3.

Table 3. Data source.

Variable	Description	Unit of Measurement	Data Source
RWE (m^3)	Raw Water Exccess (Formula (1))	m^3	calculation
RW (m^3)	Raw Water collected by the municipality: (TAC(m^3)) × (1 + percentage non accounted for water losses)	m^3	calculation
TAC (m^3)	Typical Average monthly consumption of drinking water of a household: It is calculated through TAC(USD) and the tariff specifications.	m^3	calculation
TAC (USD)	Typical Average monthly consumption of drinking water in a household in USD $	USD $	National Survey of Employment and Underemployment section environment December 2016 in Ecuador [16]
Tariff specifications	Information on the tariff structure of the drinking water service	USD $ per m^3 in blocks consumption	Available on the official pages of each subnational governments (case of Cuenca y Gualaceo), and ARCA [a] (case of Suscal)
Percentage non accounted for water losses	Percentage not accounted for water losses in the network	Percentage	Official information of ARCA
MARW	Minimum Amount of Raw Water monthly per household: MARW monthly in m^3 per habitant × number of people per household	m^3	calculation
Number of people per household	Average people in a household	People	Population and Housing Census 2010
MARW per habitant	Minimum Vital Consumption monthly per habitant in m^3: (MARW daily per habitant in l × 30 days)/1000	m^3	calculation
MARW daily per habitant in liters	Minimum Amount of Raw Water daily per habitant in liters	liters	Memorando Nro. SENAGUA-SAPYS.2-2016-0214-M
RWE(USD)	Raw Water Exccess in USD $ (Formula (2))	USD $	calculation
TAC(% AI)	Typical Average Consumption monthly of a household (Formula (3))	Percentage	calculation
TACPC(USD)	Typical Average Consumption per cápita: TAC(USD)/number of people per household	USD $	calculation
AIPC(USD)	Average Income per capita: AI(USD)/number of people per household	USD $	calculation
AI(USD)	Average income monthly of a household living in extreme poverty	USD $	National Survey of Employment and Underemployment complete section December 2016 in Ecuador [16]
TCPC(%IP)	Total cost per capita (Formula (4))	Percentage	calculation
RWEPC(USD)	Raw Water Consumption Excess Per Capita: RWE(USD)/number of people per household	USD $	calculation
IH	Impact on household (Formula (5)): growth monthly in payment for drinking water monthly consumption	Percentage	calculation

[a] Water regulation and control agency (ARCA by its initials in Spanish).

4.3. MARW Impact Process: Conceptual Framework

Figure 3 shows the impact scheme. During the process by which raw water is collected by service providers to convert it into drinking water for human consumption and its distribution to households, operators face unaccounted-for water losses (Unaccounted-for water losses, according to the Regulatory and Control Agency, are measured by dividing the average monthly commercialized water, and the average monthly water distributed to the network). In this sense, the possible scenarios are as follows: (1) scenario in which the operator does not exceed the value of wholesale and therefore converts 100% of raw water into drinking water; (2) scenario where the operator has a percentage that exceeds the value of increase, so that the operator needs more than 200 LPCPD to provide 200 LPCPD of drinking water to a home. In this context, operators who exceed the minimum vital amount of raw water to supply households, incur a cost of USD $0.0039 per m³, a cost that is not covered by those who are efficient enough to not generate excess.

In the case of interest, in which excess cost is generated, the value can be assumed by the operator without transferring it to the consumer, which increases the rate if the household maintains typical or higher levels of consumption. Therefore, the following would occur.

4.3.1. Less Efficient Operators Who Assume the Cost of the Excess

Those water service providers who face high loss percentages, have higher operating costs; therefore, if they assume the excess costs, the entity is affected in its solvency, which in turn affects its investment capacity to make improvements in the system.

4.3.2. Less Efficient Operators Who Charge the Excess Cost through the Rate Paid by the Household

Operators with high percentages of losses that index the costs of excess into rates, cause an increase in them. As a result, if a household chooses to maintain its typical consumption, it would have to pay more for the liquid and its income would be affected. However, it may also choose not to affect its income and would then be forced to reduce consumption levels which consequently affect life quality (see Figure 3).

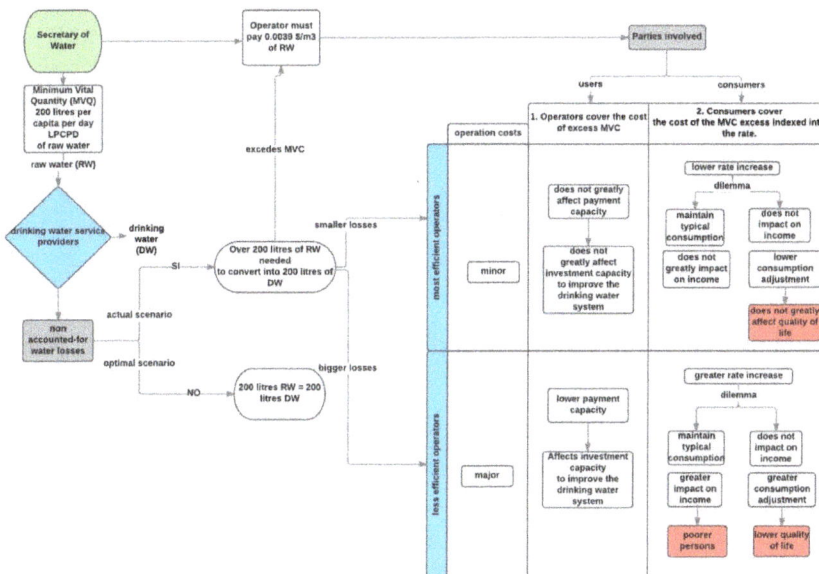

Figure 3. MARW impact.

5. Results

5.1. Economic Analysis

5.1.1. Typical Consumption of Water by Local Governments

Local governments need raw water to transform it into drinking water. The level of raw water required depends on the level of efficiency at all stages of service delivery. In this case, it is measured by the percentage level of unaccounted-for water losses. Figure 4 shows the typical household consumption of drinking water by local governments, and it also shows the raw water needed by a local government to supply the consumption of drinking water to the household in question.

In this sense, in the city of Cuenca, a household consumes on average 27.06 m^3 of drinking water per month. However, the water service operator of the city needs an average of 34.82 m^3 raw water per month to supply the usual consumption of a home, due to a 28.66% of unaccounted water losses. In the case of Gualaceo, the quantities of water needed to supply the current water consumption are even higher (39.24 m^3 of raw water to supply 25.85 m^3 of drinking water as the average household consumption), due to a higher percentage of losses (51.78%). In Suscal, the operator of this city needs on average 62.93 m^3 of raw water per month to supply 40.24 m^3 of drinking water as the usual average consumption of one household per month. The high consumption of this city's households may be due to the low prices paid by households (on average \$3.7) and the likely use of water for other activities not included in the MARW calculation formula.

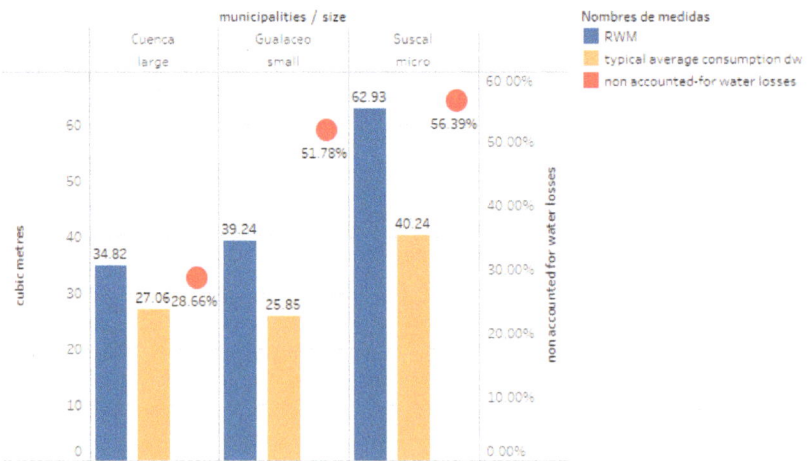

Figure 4. Typical water consumption of a household by local governments.

Using Equation (1), the results indicate that the households that consume larger amounts of drinking water require, in proportion, larger amounts of raw water; an amount which is also higher in cantons where the percentage of losses is higher. In Cuenca (canton with a lower percentage of losses compared to the rest of the cantons under study), RWE is 12.67 m^3. In addition, RWE are even larger in small and micro-sized cantons such as Gualaceo (16.70 m^3) and Suscal (39.56 m^3), which is evidently due to the high average consumption of a household (as in Suscal) and the high percentages of losses (see Figure 5a).

As a result, applying Equation (2), we obtain that the operators with lower rates of losses (Cuenca's case) would pay the state about \$0.05 monthly per household, if Cuenca household maintains its usual average monthly consumption levels of drinking water. The most affected operators are those with the higher rates of losses and higher consumption per household. Therefore, if Gualaceo households continue with the usual average consumption, the operator should pay the state $0.06 monthly per household, while Suscal must pay a value around $0.15 per month per household, which is about 3 times more than what a more efficient operator (such as Cuenca) should pay (see Figure 5b).

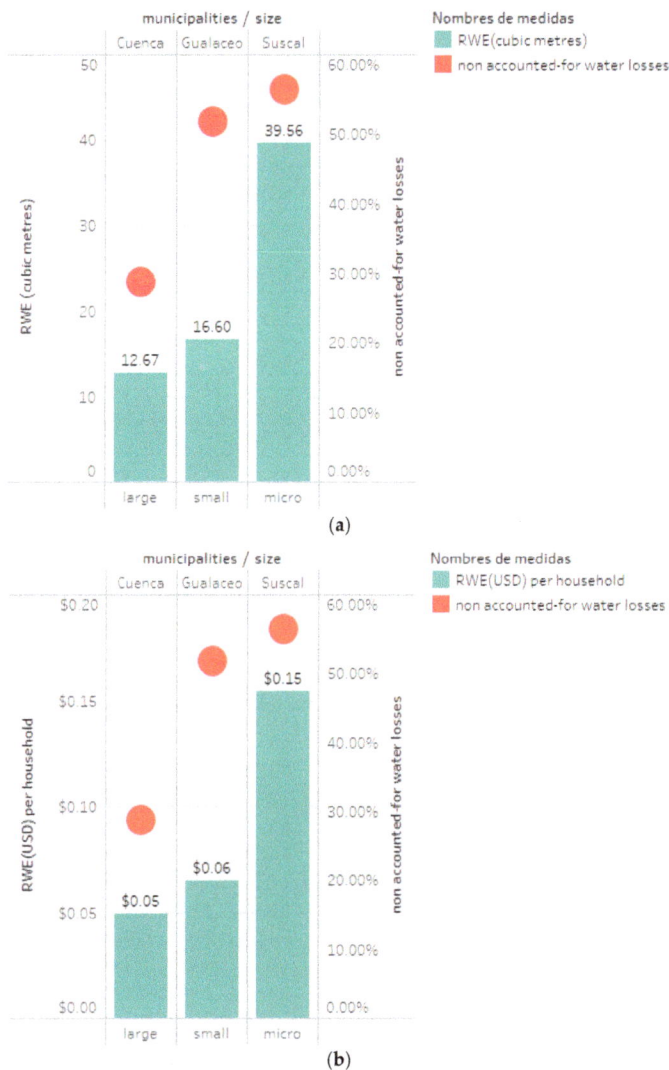

(a)

(b)

Figure 5. Excess use of raw water in m^3 and USD, according to the percentage of unaccounted-for water losses of a household with a typical level of consumption, per local governments. (**a**) Excess use of raw water in cubic meters of a household according to the percentage of water losses not accounted-for by local governments; (**b**) Excess use of raw water in monetary units of a household according to percentage of water losses not accounted-for by local governments.

5.1.2. MARW Impact on Households Living in Extreme Poverty by Income

Formulas (3)–(5) were applied to measure the effects of the application of the vital minimum of water consumption on households considered to be extremely poor by income. To do so, the average number of persons per household is considered, following the 2010 census of population and housing in Ecuador, which is 3.69; 3.77 and 3.9 for the cantons of Cuenca, Gualaceo and Suscal, respectively. This means that the minimum vital monthly amount of water for a household is 22.1; 22.6 and 23.4 cubic meters of drinking water for the same cantons, in that order.

Since there are different levels of unaccounted-for water losses, there is an excess of raw water use compared to the average monthly water consumption of a household. If the excess MARW rate (USD $0.0039) is applied to the excess MARW used, the monthly cost per household for excess consumption is of $0.05; $0.06 and $0.15 per local governments respectively (see Table 4), when the local governments rates the excess. If it does not, it ought to use other sources of income to revert the payment to the State. In the cases in which the operators index the payment for the excess into the rates, the household should pay for the excess. Consequently, those households living in poverty would be mostly affected.

Table 4. Excess consumption of minimum vital raw water in a household in m^3 and monetary units per local governments.

Municipalities	Size	Average Persons Per Household	HC Monthly Average (USD/Household)	RWE (USD)	AIPC (USD)	TAC (% AI)	TCPC (% AI)	IH
Cuenca	large	3.69	$15.78	$0.05	$30.42	14.05%	14.10%	0.31%
Gualaceo	small	3.77	$8.02	$0.06	$34.50	6.16%	6.21%	0.81%
Suscal	micro	3.90	$3.70	$0.15	$27.97	3.40%	3.54%	4.17%

After applying the Formulas (3)–(5), the typical consumption per capita is obtained as a percentage of the average monthly income per capita in households considered extremely poor. By including the cost of excess MARW, the total cost to be paid by the household for the consumption of raw water increases in relative terms. The generated variation represents the affectation to the home considered in extreme poverty, thus, a home belonging to canton Suscal should pay an additional 4.17% (of their average income). The results indicate that households with greater economic difficulties and also those in cities with less efficient operators, experience a greater impact on their income (in the case of households in Suscal) compared to households considered in extreme poverty living in cities with more efficient water service providers (there are less losses in the case of Cuenca).

In sum, the application of the minimum vital water consumption has economic implications both for service operators and the families that belong to these municipalities. The level of inefficiency (efficiency) is a key factor in the final outcome of the implementation of the MARW. Thus, operators with less water losses will face lower costs (or none, if their loss is within the increase value of the formula) and therefore the households they serve can maintain their typical levels of consumption and do not see their income affected. On the other side, operators that handle high levels of raw water loss will assume higher costs to provide the service, while households face the dilemma of maintaining their levels of consumption and pay more, if the operator charges the cost, or modifying their levels of consumption with the subsequent effect on their quality of life.

Results show that, in fact, there is a correlation between the size of the local governments, the efficiency of the operator, and the affectation to the income of the poorest families; which implies that the application of the MARW adversely affects the smaller municipalities, the less efficient operators, and the most economically vulnerable families.

6. Conclusions

The standardization of MARW in Ecuador does not consider differences in municipalizes' size, efficiency in service provision and income levels of their populations. Far from being a convergence factor, it widens the differences between them. Regarding water services provision, the enforcement of vital minimum amount of water regulation "punishes" least efficient operators. Therefore, to sustain their finances, operators transfer to the cost to their clients, that is, the families which are provided with the service. The families most affected by the application of the MARW are those settled in municipalities with lower incomes and a greater indigenous population. The possibilities for these families are to pay more in order to maintain level of consumption or consume less in order not to adversely affect the family finances, even if this means affecting their quality of life. These findings suggest that the water authority should establish an MARW that captures the differences between the Ecuadorian municipalities.

Acknowledgments: The authors thank the research assistance of Eco. Fanny Cabrera from the Faculty of Economics and Administration of the University of Cuenca, and students Adriana Abril and Francisco Bermeo from the school of law at the Faculty of Jurisprudence, Political and Social Sciences of the University of Cuenca.

Author Contributions: Andrés Martínez Moscoso designed Section 2. Background and Section 3.2. The vital minimum Quantity of Drinking Water in the Ecuadorian Legislation; Víctor Gerardo Aguilar Feijó designed Section 4. Methodology and Section 5. Results; and Teodoro Verdugo Silva contributed with the Section 3.1. Analysis of International and National Case Law on the Vital Minimum Amount of Drinking Water.

Conflicts of Interest: The authors declare no conflict of interest.

References and Note

1. Pigretti, E.; Dino Bellorio, C.; Cavalli, L. *Environmental Law of Waters*, 1st ed.; LAJOUANE, Gráfica Sur Editora: Buenos Aires, Argentina, 2010; ISBN 978-84-7429-522-1.

2. Decentralized Atonomous Municipal Government of Canton Cuenca. Cuenca Has the Best Drinking Water in the Country. Available online: http://www.cuenca.gob.ec/?q=content/cuenca-cuenta-con-la-mejor-agua-potable-del-pais (accessed on 8 December 2017).

3. Instituto Nacional de Estadística y Censos (INEC). Population Projections. Available online: http://www.ecuadorencifras.gob.ec/proyecciones-poblacionales/ (accessed on 8 December 2017).

4. Instituto Nacional de Estadística y Censos (INEC). Population and Demography. Available online: http://www.ecuadorencifras.gob.ec/censo-de-poblacion-y-vivienda/ (accessed on 8 December 2017).

5. World Health Organization/United Nations International Children's Emergency Fund. *WHO/UNICEF Joint Monitoring Program for Water Supply and Sanitation*; Global Water Supply and Sanitation Assessment 2000 Report; WHO/UNICEF: Washington, DC, USA, 2000.

6. Thompson, J.; Porras, I.T.; Tumwine, J.K.; Mujwahuzi, M.R.; Katui-Katua, M.; Johnstone, N.; Wood, L. *Drawers of Water II: 30 Years of Change in Domestic Water Use and Environmental Health in East Africa*; International Institute for Environment and Development: London, UK, 2001.

7. Chenoweth, J. Minium water requirement for social and ecomomic development. *Desalination* **2008**, *229*, 245–256. [CrossRef]

8. Scanlon, J.; Cassar, A.; Nemes, N. *Water as a Human Right*; IUCN Envoronmental Law Program; International Union for Conservation of Nature: Gland, Switzerland, 2004.

9. Embid-Irujo, A.; Domínguez-Serrano, J. (Dir.). *The Quality of Water and Its Legal Regulation. (A Comparative Study of the Situation in Spain and Mexico)*; Iustel, Portal Derecho SA: Madrid, Spain, 2011.

10. De Albuquerque, C. *Report of the Special Rapporteur on the Human Right to Drinking Water and Sanitation*; United Nations: Geneva, Switzerland, 2013.

11. De Albuquerque, C. *Realising the Human Rights to Water and Sanitation: A Handbook by the UN Special Rapporteur Catarina de Albuquerque*; Office of the High Commissioner for Human Rights, UN Habitat: Bangalore, India, 2014.

12. Instituto Nacional de Estadística y Censos (INEC). Census of Population and Housing. Available online: http://www.ecuadorencifras.gob.ec/sistema-integrado-de-consultas-redatam/ (accessed on 8 December 2017).

13. Centro de Investigaciones Sociales del Milenio (CISMIL). II National Report of the Millennium Development Objectives Ecuador 2007. Partnerships for Development. Available online: http://www.undp.org/content/dam/ecuador/pnud_ec_II_INFORME_NACIONAL.pdf (accessed on 8 December 2017).

14. Corte Constitucional de Colombia. Available online: http://www.corteconstitucional.gov.co/relatoria/2011/T-740-11.htm (accessed on 17 October 2017).

15. Constitutional Court of Ecuador. SENTENCE No. 001-12-SIC-CC, CASE No. 0008-10-IC. Available online: http://portal.corteconstitucional.gob.ec:8494/FichaRelatoria.aspx?numdocumento=001-12-SIC-CC (accessed on 15 December 2017).

16. Ecuador, Organic Code of Production, Trade and Investment, Art. 94. Available online: http://www.wipo.int/wipolex/es/details.jsp?id=11417 (accessed on 25 October 2017).

17. SENAGUA. Memorandum No. SENAGUA-SAPYS.2-2016-0214-M. 2016.
18. Instituto Nacional de Estadística y Censos (INEC). National Survey of Employment, Unemployment and Underemployment. Indicators of Poverty and Inequality. 2016. Available online: http://www.ecuadorencifras.gob.ec/empleo-diciembre-2017/ (accessed on 20 October 2017).

resources

MDPI

Article

From Fragmented to Joint Responsibilities: Barriers and Opportunities for Adaptive Water Quality Governance in California's Urban-Agricultural Interface

Ann Drevno

Department of Environmental Studies & Sciences, Santa Clara University, Santa Clara, CA 95053, USA; adrevno@scu.edu

Received: 4 December 2017; Accepted: 7 March 2018; Published: 17 March 2018

Abstract: California is facing a critical water supply and water quality crisis, necessitating a clear shift in the way water resources are managed. This study assesses the effectiveness of water law and policy in the urban-agricultural interface, where the two discharge into common waterways but have different regulatory requirements. A case study from one of California's most productive agricultural regions, the Salinas Valley, explores the complexities and inadequacies of current water law in the interface, as well as promising integrated water management schemes. The article's findings are based on archival research, extensive document review and 15 in-depth interviews with key stakeholders. Findings suggest that local, state and federal water policy is severely fragmented, providing little incentive for the multitude of water entities to collaborate on multi-benefit projects and resulting in unsuccessful water quality improvements. There is a strong need for a more integrated policy approach that bridges different types of dischargers (agricultural and urban), water quality and water quantity issues and also incorporates land uses into policy decision making.

Keywords: water law and policy; water resource management; urban-agricultural interface; California

---

1. Introduction

California is facing a critical water supply and water quality crisis, necessitating a clear shift in the way water resources are managed. Clean water is vital to human health, ecosystem functions and the economy and has never before been in such high demand. In California, water pollution was exacerbated by the most severe drought on record followed by El Niño rains [1]. During the unprecedented five-year dry spell, the state lost roughly 11 trillion gallons of water [2], resulting in the literal sinking of land [3]. Of the limited water the state did have, the vast majority—90% of all assessed waters—had some degree of contamination [4]. While the 2016–2017 winter, the wettest on record, pulled the state out of the drought and ended the drought state of emergency across the entire state, it aggravated water quality issues by increasing water runoff and accompanying contaminants. Additionally, more frequent and severe weather events due to climate change will continue to have a significant impact on water resources [5].

The brunt of health problems related to water pollution has fallen on the most vulnerable and marginalized populations, with nitrates and pesticides being two primary health constituents of concern in California. Over 2 million people in the state, mostly low-income, minority farmworkers, are at risk of drinking nitrate-contaminated water due in large part to agricultural pollution [6]. Schools in the Central Valley's farmland have found such high concentrations of pollutants that they have cut off their drinking fountains to students. Nitrate-contaminated drinking water from agricultural fertilizers is a well-known risk factor for "blue baby syndrome," a potentially fatal blood disorder resulting in reduced oxygen-carrying capacity of hemoglobin [7]. Because these communities are among the poorest in California, many lack the resources or technical capacity to maintain safe drinking water supplies [6,8], producing a startling water insecurity problem in the country's richest state. Inequitable distribution of agricultural water contamination and its consequences have become top public policy concerns. Yet there is a dearth of studies that have addressed how the existing legal water doctrines have failed to meet basic water quality standards and health protections.

One of the most suitable places to observe this water quality management challenge is in the state's expanding urban-agricultural interface, where exposure to agricultural water pollution is aggravated by sheer proximity and where jurisdiction of common waterways overlap. Urban areas are not without their pollution problems. However, urban runoff is regulated much more stringently than agricultural runoff, creating a rich opportunity to compare and contrast urban and agricultural pollution control management strategies and the emergence of alternative multi-level legal arrangements to clean conjoint waters. These discrepancies create the conditions for a unique "natural experiment" to examine the different and uneven institutional arrangements and implementation practices governing water resources between urban and agricultural sectors. Additionally, the demand for improved water quality in the urban-agricultural interface presents an opportunity to learn from emerging water pollution governance strategies that may allocate benefits more equitably and modify existing legal regimes to adapt to new contexts.

In the urban-agricultural interface, institutions charged with protecting the state's water resources are numerous and fragmented. Urban water quality is generally regarded as a municipal issue and agricultural water pollution, a nonpoint source, is exempt from discharge permitting and shrugged off as near- impossible to regulate due to its diffuse nature.

As the interface expands and pollution flows into different jurisdictions, municipalities, together with agencies that regulate agricultural operators, are forced to devise management plans to comply with federal and state water quality standards. While a variety of management options are available to these newly formed partnerships, the task of selecting and implementing appropriate policies and laws is challenging since the two sectors often have conflicting interests, priorities and socio-hydrologic contexts. Even though California is often viewed as a pioneer in environmental policies, the state's multitude of stakeholders, including a powerful agricultural lobby, often makes California water quality decision-making fraught with tension.

This research fills a much-needed gap of examining the barriers and opportunities for governing water quality in some of California's most vulnerable and politicized landscapes. This paper is organized into five parts. After a brief description of research methodologies, the paper: (1) describes three pollutants—nutrients, sediments and pesticides—and the harm they inflict on water bodies in urban and agricultural waterways, especially when they accumulate; (2) reviews several key water quality regulations in urban and agricultural land use areas, including three adaptive legislative measures related to state water issues; (3) provides a case study that exemplifies the complexities and inadequacies of current water law in the urban-agricultural interface, as well as promising integrated water management schemes; and (4) concludes by recommending avenues for advancing more coordinated and effective responses to water quality management.

2. Research Design and Methods

The article's findings are based on archival research, extensive document review and 15 in-depth interviews with water resource managers, engineers, city planners and officials, non-governmental representatives and policymakers. Each key informant was personally or professionally involved with and knowledgeable about historic and/or current water governance systems. Semi-structured interviews included topics such as: What factors motivated the regional, state and federal authorities to draft distinctly different rules for water pollution in urban and agricultural areas? What were the goals when creating these protections, or exemptions? Have water quality governance structures evolved and strengthened over time (through amendments, rulings or revisions) and why? Have there been any efforts to restructure ineffective, disjointed water management strategies and move toward a more collaborative, multiparty approach to improving water quality in conjoint waters? Research was also based on participant observation at conferences and workshops devoted to agricultural water quality and storm water control, as well as review of policy documents and reports.

3. The Cumulative Toxic Brew in the Urban-Agriculture Interface

The water quality problems posed by nonpoint sources are numerous and diffuse in the urban-agricultural interface. When multiple inputs from urban storm drains and runoff from farms accumulate in a single waterbody, the impacts are magnified, causing cumulative effects [9]. Agricultural and urban storm water runoff are two of the top sources of waterbody impairments in the state [4]. Approximately 26,261 miles of streams and rivers and 172,050 acres of lakes are impaired by agriculture [4], while urban storm water likely contributes to 124,557 acres of impairments in lakes. Construction, a predominantly urban activity, accounts for another 88,850 acres of impairments in lakes and an additional 15,469 miles of impairments in California rivers and streams [4].

The three predominant agricultural nonpoint sources of pollution—nutrients from fertilizers, pesticides and sediments from soil erosion—together are the chief impediments to achieving national water quality objectives [10]. A UC Davis report commissioned by the California State Water Board reveals high levels of nitrate contamination in the Tulare Lake Basin and Salinas Valley predominantly from over-application of fertilizers in agricultural areas [6]. Pesticides are another agricultural contaminant of concern due to their more obscure impact on human health and the environment than their nutrient and sediment pollutant counterparts [11]. Pesticide use in California is known to contribute to water column and sediment toxicity [12–14], as well as cause human health problems, such as developmental delays in infants and children [15].

Urban storm water also contains an assortment of pollutants including nutrients and pesticides from landscaping, siltation from development projects and chemicals, oily residue and salt from impervious city surfaces [11]. Urbanization has compromised water quality through pollutant loading and by increasing stream temperatures through reducing shade and converting natural vegetation to impervious surfaces; these activities have negatively affected fish and invertebrate populations [16]. Conventional urban drainage systems often channel runoff directly to nearby waterways, thus exacerbating pollutant inputs and ecological disturbance [17]. During rain events sewage treatment systems can overflow, releasing raw sewage from the collection system before reaching the treatment facility [18]. Trash pollution has long plagued urban water quality regulators and has recently come to the forefront of urban pollution issues due to the 2015 adoption of statewide "Trash Amendments", which require all municipalities to install trash capture devices or control technologies in priority outfalls throughout their jurisdictional boundaries. Because it is largely an unfunded mandate, city managers are wondering how and where they will find the funds to implement the expensive new infrastructure requirements.

Combining individual discharges from agriculture and those from urban areas can often make a significant, adverse change to the water quality in the interface [9]. For example, if a farmer over applies fertilizer on a crop and an urban landscaper did the same on a lawn within the same watershed, the excess nutrients from each event could multiply, stressing the receiving waterways and/or putting the ecosystem at risk of pollution [9].

4. Divergent Laws in Merging Waters

Policies that regulate discharges from cities are distinctly different than those that regulate discharges from agriculture. While both urban and agricultural industries pollute to common waters, the legal structures that manage those discharges are different. Because cities are considered point source polluters and regulated by discharge limitation permits, municipal laws aimed at water pollution are significantly more stringent and different from agricultural ones. Discharges from irrigated lands (farms) are considered nonpoint sources and, in California, regulated through Conditional Agricultural

Waivers. Water pollution control technologies, such as the implementation of Green Infrastructure programs, are also highly variable between urban and agricultural areas. The following describes the principal regulations affecting water quality in the urban and agricultural landscapes. Table 1 lists a more comprehensive set of water quality policies on the local, state and federal level.

The 1972 U.S. Clean Water Act (CWA) was established to protect the waters of the United States. The CWA regulates the discharge of pollutants into a local water body by setting uniform numeric discharge limits. These effluent limits are calculated based on the type of industry that is discharging as well as the beneficial use(s) (i.e., drinking, swimming, fishing, etc.) of the receiving water body. Restrictions are enforced through National Pollution Discharge Elimination System (NPDES) permits and apply to all point source polluters. Point sources of discharge are defined in the legislation as "any discernible, confined and discrete conveyance, including … any pipe, ditch and channel" (CWA § 502). Besides Agricultural Feeding Operations (AFOs), this definition excludes agricultural discharges, considering them a nonpoint source of pollution, thereby requiring states to develop ways of controlling them.

The U.S. EPA delegates to most states, including California, the authority to administer and enforce its own NPDES permits to point source dischargers. California's comprehensive 1969 Porter-Cologne Act established the State Water Resources Control Board (SWRCB), or "State Water Board" and gave broad authority to nine Regional Water Quality Control Boards, or "Regional Boards," to regulate water quality at a local level. The Regional Board's authority includes issuing and enforcing all NPDES permits, as well as waiving those permits for certain polluting industries. All Regional Boards have chosen to waive waste discharge requirements from agriculture by employing what is called "Conditional Agricultural Waivers" or "Orders."

Table 1. Federal and state water quality regulations and associated goals.

Scale of Governance	Agency	Goal
Federal	1972 Clean Water Act	Regulates water pollution; adopts water quality standards
	Safe Drinking Water Act	Establishes drinking water standards for contaminants that may cause health effects
	Endangered Species Act	Prevent extinction, recover imperiled species
State (California)	Porter Cologne Act	Comprehensive program to protects water quality in California and the beneficial uses of water
	Sustainable Groundwater Management Act *	Established long-term, local groundwater management
	Human Right to Water Act *	Establishes the human right to safe, affordable, clean and accessible water
	Affordable Right to Water Bill (Proposed) *	A Bill proposed to provide financial assistance to communities that lack safe drinking water
Regional (within California)	Conditional Agricultural Waiver	Water pollution control from irrigated lands
	Basin Plan	Acts as a master quality control planning document, setting beneficial uses and water quality objectives
	MS4 General Permits	Water pollution control from municipal runoff

* Legislation that necessitates collaboration between urban and agricultural water quality agencies (described below).

This approach has not gone uncontested. The program requires Boards to attach conditions to waivers and review them every five years. In a recent Central Coast Conditional Agricultural renewal, scientists, the California Department of Public Health, environmental justice groups and environmentalists called into question the effectiveness of this more lenient form of regulation in protecting water quality. A particularly frustrated coalition of environmental groups, together with an elderly woman who could not drink water from her tap due to agricultural contamination, filed and won a lawsuit in Sacramento's Superior Court challenging the legitimacy of the Central Coast Regional Conditional Agricultural Waiver. The coalition claimed that the Ag Order was "so weak it did not comply with state law" [18]. In his ruling on 11 August 2015, Superior Court Judge Frawley, agreed that the Central Coast's Conditional Agricultural Waiver was doing an inadequate job of protecting regional water quality and needed to develop more stringent conditions [19].

While agricultural nonpoint discharges are conditionally waived, urban sources of water pollution, including urban storm water, are regulated as point sources through NPDES permits, as described above. The 1987 amendments to the CWA broadened the definition of "point sources" to include municipal and industrial storm water runoff, adding section 402(p) to the Clean Water Act. In subsequent years, the EPA developed the municipal separate storm water systems (MS4) program in two phases (Phase I: 1990; Phase II: 1999). The 1990 Phase I regulation required medium and large cities or certain counties with populations of 100,000 or more to obtain MS4 NPDES permit coverage for their storm water discharges. There are approximately 855 Phase I MS4s covered by 250 Individual Permits across the country. In 1999, the Phase II regulation established storm water control mandates for small cities with a population of fewer than 100,000. Rather than issue individual storm water permits for every municipal discharger, the California State Water Resources Control Board, as authorized by the Phase II rule, adopted a "General Permit" for all small the Municipal Separate Storm Sewer Systems (MS4s), which include all storm drains that discharge into local waterbodies (Order No. 2003-0005-DWQ). As such, municipal storm water systems obtain coverage under the general permit from their Regional Board.

There are several types of permits, which differ in their requirements. The requirements of the *MS4* general permit include developing and implementing a comprehensive Storm Water Management Program (SWMP), outlining management practices to reduce pollutant discharges. An additional requirement of the general permit is that local governments must also inspect construction sites and industrial facilities for compliance with the *industrial* general NPDES permit (which includes construction activities). The general permit has its own set of requirements, including a separate plan, called the storm water pollution prevention plan (or "SWPPP"), detailing how and when the responsible party will implement erosion and sediment control among other Best Management Practices (BMPs). This involved set of requirements and permits means that urban storm water (sewers, construction sites, commercial and industrial facilities, etc.) are regulated by the U.S. EPA, the California EPA, the Regional Water Quality Board and local governments. Management of these complex water systems and laws can be very challenging due to the multitude of separate entities without much coordination [20]. Table 2 lists the major agencies charged with water quality in California and their associated responsibilities.

Table 2. Major water agencies and their associated roles and responsibilities.

Scale of Governance	Agency	Responsibilities
Federal	U.S. EPA	Regulates water quality through the Clean Water Act, Safe Drinking Water Act and other laws
State	Water Resources Control Board Department of Water Resources Department of Public Health Department of Pesticide Regulation Public Utilities Commission	Implement CWA provisions; administers state water rights Oversees state water planning Regulates drinking water Regulates statewide pesticide use Regulates water rate structures for private utilities
Local	Regional Water Quality Control Boards Agricultural Commissioners Offices Natural Resources Conservation Districts County Environmental Health Department Municipal governments	Regulates water quality Local administration of pesticide use enforcement Local financial and technical assistance to farmers Local administration of domestic water systems Local administration of MS4 General Permit

Note: List not exhaustive.

The discrepancies between urban and agricultural water quality regulations have not gone unnoticed. Facing increasingly stringent urban storm water pollution control regulations, municipalities have begun to question the fairness of waiving discharges from agriculture. Municipalities have voiced their concerns about pollutants from agricultural areas being deposited into receiving waterbodies within city boundaries. One former City Manager, Fred Meurer of Monterey, suggested that agricultural industries and cities should be held to the same high water quality standards [21].

One mechanism California has employed to improve water quality conditions in complex landscapes is through the proposal and passage of Legislative Bills aimed at particularly acute water issues, such as the right to clean drinking water. While these bills do not directly address agriculture's relatively lax water quality regulations or the fragmentation that exists between agricultural and urban water agencies, they have forced more collaboration between agencies on these issue-based mandates. The following section examines three legal protections—the Sustainable Groundwater Management Act, the Human Right to Water Act and the Affordable Drinking Water Bill—all separate bills aimed at addressing specific water quality issues in California.

5. Adaptive Legal Frameworks

In geographic areas where jurisdiction overlaps and health hazards are evident, new multi-level institutional arrangements are being developed and implemented to address water quality challenges [22]. Each of the following examples offers valuable insights in both the water pollution issues (i.e., human health, affordability, access) and challenges (i.e., multitude of different actors, who will pay) faced in the growing urban-agricultural interface.

5.1. Sustainable Groundwater Management Act

For the first time in state history, the Sustainable Groundwater Management Act (SGMA) established a framework for long-term, local groundwater management. Passed in 2014, this was a multi-faceted three-bill package that makes progress in restricting groundwater over-extraction. In the state's high and medium-priority groundwater basins, which account for 96% of the state's groundwater use, local water districts are now required to bring their basins into balanced levels of pumping and recharge. The Act identifies six categories of actions that could have an "undesirable result" on sustainable groundwater management, including: persistent lowering of groundwater levels, significant and unreasonable reductions in groundwater storage, significant and unreasonable salt water intrusion, significant and unreasonable degradation of water quality, significant and unreasonable land subsidence and surface water depletion having significant and unreasonable effects on beneficial uses. While two goals focuses on water quality, because SGMA was a drought-driven piece of legislation the prime focus has been on water supply/quantity. SGMA established the formation of locally-controlled groundwater sustainability agencies (GSAs) to adopt management plans tailored to their community's circumstances and needs. The strategy of delegating control to local agencies. However, some question whether these newly-founded GSAs will have sufficient regulatory authority to be effective.

5.2. Human Right to Water Act

In 2012, California became the first state to legally recognize the human right to water [8]. The Human Right to Water Act recognizes that "every human being has the right to safe, clean, affordable and accessible water adequate for human consumption, cooking and sanitary purposes". Under the new law, the California Water Code now requires all relevant agencies, specifically the Department of Water Resources, the State Water Board and the California Department of Public Health, to *consider* the human right to water when making policy decisions. The duty to consider includes taking into account several substantive factors—quality, quantity, affordability and accessibility—that may impact access to safe water.

California's human right to water policy was born out of a devastating lack of clean drinking water for numerous communities, the most pervasive of which are in disadvantaged rural areas with agricultural runoff [8]. The State Water Board estimates that roughly 300 disadvantaged communities in the State receive water from public systems that do not meet drinking water standards [8]. Because the policy only provides guidance, the bill does not create a right of action for customers to demand clean water. However, the "duty to consider" directive does create a legally binding mandate for administrative state agencies to think about how their decisions might impact drinking water quality and public water systems.

5.3. Affordable Drinking Water Bill (Proposed)

Senate Bill 623 (As of 1 September 2017, the Bill remains active and under review by the Rules Committee). addresses a gaping hole in the Human Right to Water Act—who will pay for clean water in disadvantaged communities who lack access? Many of the State's small public water systems with polluted drinking water lack the technical, managerial and financial capacity to clean and deliver safe water at affordable rates. The State has provided Proposition 1 Funds, which offers $7.5 billion to fund watershed protection and water supply projects throughout the state, as well as Drinking Water State Revolving Funds for assisting disadvantaged communities with paying for treatment systems. However, these are short-term solutions and currently no such funding mechanisms are available for longer-term operational improvements. If passed, the Bill would generate $2 billion over 15 years for the Safe and Affordable Drinking Water Fund in the State Treasury. The Fund would provide emergency water and longer-term system fixes for hundreds of communities whose tap water is not in compliance with safe drinking water standards. While most agree with the Bill's goal—assisting disadvantaged communities that lack potable water, public water agencies strongly oppose the Bill's proposed funding mechanism: taxing local water bills (95 cents per month per water customer). Another funding mechanism that would be employed is a fertilizer fee to address contamination from farms. Interestingly and despite the cost to some farmers, agricultural and environmental groups have come together in an unusual coalition to support this bill.

6. Case Study

California abounds with urban-agricultural interfaces that are well-suited for highlighting these issues. A case study from one of California's most productive agricultural regions, the Salinas Valley, was carefully selected based on the presence of polluted water bodies in adjoining agricultural and urban landscapes as well as the presence of a new joint initiatives aimed at addressing water quality.

Salinas, California

The Salinas Valley in California's Central Coast Region is known as the "salad bowl of the world" for its productive lettuce heads and mixed greens. Once a small agricultural community, the City of Salinas is now the largest in the Central Coast, boasting a population of over 160,000 residents. Within its approximately 14,400 acres, the City has become "a suburban community nestled in an agricultural setting" [23], as depicted in their City logo (Figure 1) and land use patterns (Figure 2), making it an ideal case study for this research.

Because the City is surrounded by agricultural land uses, it is closely linked to the agricultural industry. Farmworkers reside in the City and commute to work in nearby agricultural operations and agricultural products, like the Valley's famous bagged salad greens, are transported into the City for industrial processing. Additionally, farming occurs in the Carr Lake area located within City limits, illustrated in the aqua-colored bulls-eye in the center of the City in Figure 2.

Figure 1. City of Salinas logo.

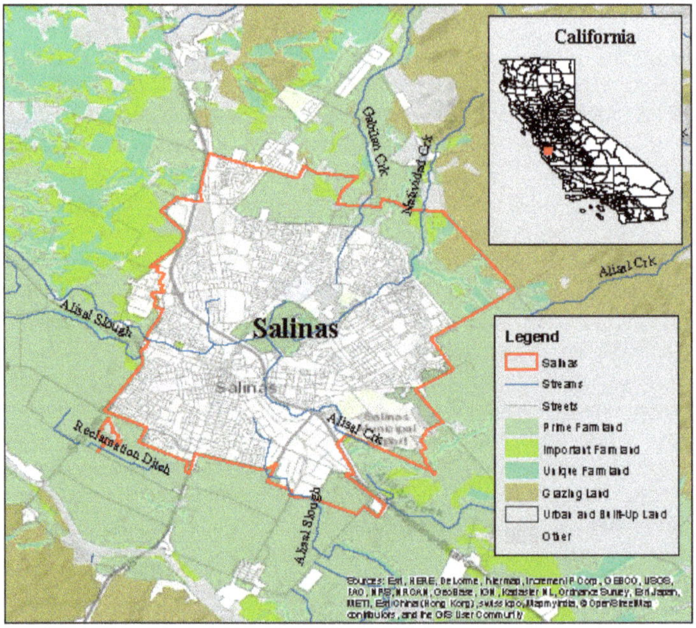

Figure 2. Salinas, CA: A city engulfed by agriculture.

Numerous waterways run through the city limits and are part of two sub watersheds that are pertinent to this case study: The Reclamation Ditch sub watershed and the Lower Salinas River watershed. The Reclamation Ditch sub watershed drains to the Old Salinas River and contains Tembladero Slough and its tributaries: The Reclamation Ditch, Espinosa Slough/Santa Rita Creek, Gabilan Creek, Natividad Creek, Alisal Creek and Towne Creek. The Lower Salinas River sub watershed drains to the Salinas River Lagoon and contains the Salinas River and its tributaries: Blanco Drain, Toro Creek, Quail Creek and Chualar Creek [24]. Both of these sub watersheds empty into the Monterey Bay. The Reclamation Ditch sub watershed, one of the most polluted systems in the region (see Figures 3 and 4), flows, on an incoming tied, into the Elkhorn Slough, a Marine Protected Area and National Estuarine Research Reserve. Nearly all of these waterways are impaired for one or a variety of pollutant parameters including ammonia, fecal coliform, low dissolved oxygen, nitrate, chloride and sodium. Figure 3 depicts a Water Quality Index calculated by the Surface Water Ambient Monitoring Program (SWAMP) for federally defined Central Coast Region watersheds, with green being good, yellow slightly impacted, red impacted and dark red severely impacted. Figure 4 shows data collected from the Central Coast Ambient Monitoring Program (CCAMP) water quality monitoring program [25], illustrating the water toxicity (based on invertebrate survival) in Salinas and nearby waterbodies.

As previously mentioned, larger cities are required to comply with a more rigorous Phase I MS4 NPDES permit. Salinas, with a population of over 100,000, is the only city in the Central Coast region that is big enough to qualify as Phase I (all other municipalities are regulated as Phase II). In 2012, the State Water Board and Central Coast Regional Board approved Salinas' most recent MS4 Phase I NPDES Permit (Order No. R3-2012-005), superseding the prior 1999 and 2004 orders. The 132-page document contains a description of regulations and best management practices (BMPs) that the City intends to employ to meet Permit requirements. When complying with its MS4 NPDES permit, the City is responsible for reducing discharges to the maximum extent practicable (MEP).

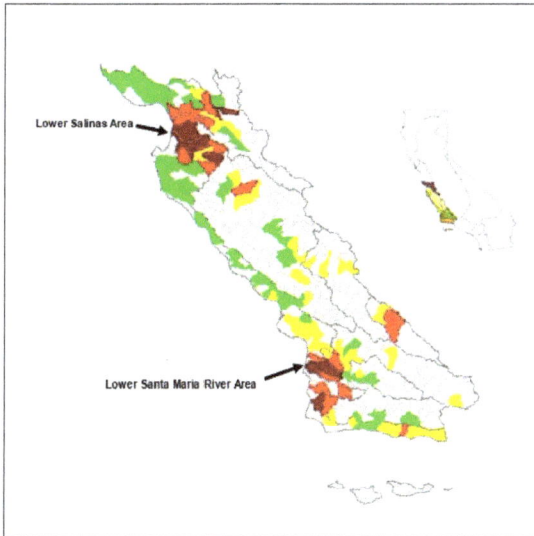

Figure 3. Water Quality in Central Coast. Source: [26].

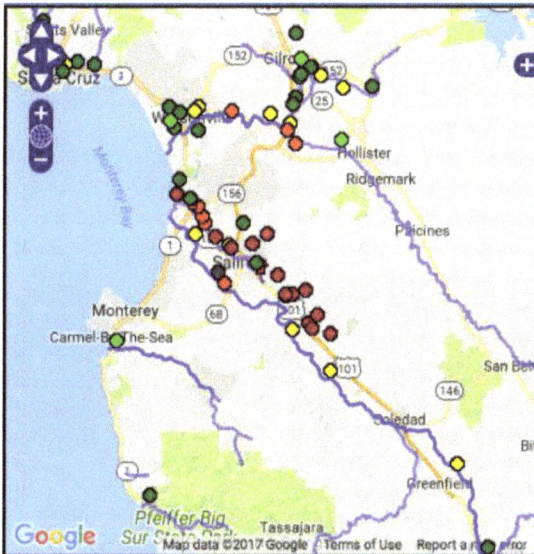

Figure 4. Water Toxicity in Salinas. Source: [25].

One of the biggest challenges urban water quality regulators face is when water entering their city is already polluted when they receive it, which, in Salinas, it usually is. Water entering into Salinas often has very high sediment and pollutant loads. According to Salinas' Stormwater Master Plan, "water flowing into the City is as much a concern for Salinas as water flowing downstream from the City. This is especially true for stakeholders furthest downstream who inherit the effects of the good, as well as the poor watershed management practices of their upstream neighbors". According to City of Salinas staff, "the major existing drainage problems occur at the boundary of the City where runoff from adjacent agricultural fields flows into the City" [23]. An early report (1994) prepared by the California Department of Fish and Game, Marine Pollution Lab and the Moss Landing Marine Laboratory, found similar pollution causes: "Agricultural lands receive higher levels of known poisons than any other landscape in the state. Year after year, farm chemicals drain into a ditch system that empties directly into the Monterey Bay Marine Sanctuary. Urban runoff is less important in the Salinas Valley than farm sources." It should be noted that while urban runoff has been deemed less of a pollution source than agriculture, it is not without its own problems, including trash, sediment, nutrient and pesticide contamination.

Because the City is responsible for the quality of water leaving its boundaries, it begs the question: is the City responsible for cleaning the polluted agricultural water it inherited from further upstream? A statement from the Stormwater Management Plan Update (2013) explains the nuanced difference in what it is and is not responsible for:

> "As operator of the MS4, the Permittee cannot passively receive and discharge pollutants from third parties. By providing free and open access to an MS4 that conveys discharges to waters of the U.S., the Permittee essentially accepts responsibility for discharges into the MS4 that it does not prohibit or control. These discharges may cause or contribute to a condition of contamination or a violation of water quality standards. However, discharges from agricultural lands that are comprised solely of return flows and/or storm water are exempt from NPDES permitting. As such, the Permittee is not responsible for these discharges that enter its MS4. The Permittee is responsible for other agricultural-related discharges into its MS4."

In summary, Salinas cannot passively receive upstream discharge pollutants, unless they are from agriculture, in which case they can acquiescently allow agricultural contaminants to flow through their city's waterways. Lacking little to no control of the agricultural pollutants that come into the City's jurisdiction and little incentive or regulatory authority to clean them up, the situation leaves what one water manager describes as "futile" (Personal Communication, 15 November 2017). The intricate hydrologic structure in and out of the City further complicates the variance in urban and agricultural regulatory systems. The following description depicts the convoluted physical flow of water and tangled assemblage of urban and agricultural dischargers in Salinas' waterways:

> "Water that begins its journey in the relatively undisturbed Gabilan and Santa Lucia Mountains drains farmlands and other cities and developed areas before entering Salinas. Once in the City, water passes through municipal neighborhoods before re-entering farmlands, then flows on to more urban uses. Water flows out of Salinas to re-enter more farmland before draining ultimately to Monterey Bay. On its journey, water flows through several different land uses, some more than once and often through several different jurisdictions" [23].

If the City of Salinas is not required to clean up agricultural storm water but urban and agricultural discharges mix into a complex stew of amalgamated pollutants, how do regulators know which pollutants are from agricultural lands and which are from urban and other land uses? The short answer is: they do not. Leaving the unknown sources of pollutants subject to the blame game of different stakeholders pointing fingers at one another defending that the problem is not theirs to clean up. However, with the emergence of several listings of impaired waterbodies in the area, efforts are underway to tease out who polluted and how much. Recent 303(d) listing in the Salinas River watershed include: chlorpyrifos and diazinon Total Maximum Daily Load (TMDL) (2011), fecal coliform TMDL (2010), nutrient TMDL (2013), salts TMDL (in development), sediment toxicity TMDL (in development) and turbidity TMDL (in development). While water quality agencies (in the city, region and state) are ramping up monitoring efforts in these contaminated hotspots to gain a better understanding of the major polluters, another key element to consider in the TMDL process is that it is not just the total amount of pollution reaching the water body but also the amount of water present to "dilute" or to "assimilate" those waters [27]. Consequently, the TMDL program "cannot work properly if water quality agencies limit pollutant discharges while the water rights department simultaneously—and without coordination—allows more water withdrawals" [27], making the chemicals in the waterbodies more concentrated.

An example from Salinas illustrates the increased need to couple water quality and water quantity when designing policies and programs. Until very recently (before the adoption of the SGMA in 2014), farmers in California have always had unlimited access to groundwater for irrigation. If they can drill a well, they can use the groundwater. In Salinas, however, that all changed a few years earlier than in the rest of the state when farmers begun sucking saltwater instead of freshwater out of their wells. In a frenzy, farmers led the charge to improve groundwater quality so they could irrigate their crops. With the dire need to mitigate saltwater intrusion (a water quality issue) and with fresh impetus from the Groundwater Sustainability Management Act (a largely water quantity-driven piece of legislation), Salinas rapidly developed more integrated approaches to connecting water quality and water quantity. One particularly noteworthy project is "Pure Water Monterey", which is a coordinated effort between the City of Salinas, Castroville, the County of Monterey, the Monterey Peninsula Water Management District and the Monterey Regional Water Pollution Control Agency ("Monterey One"). Pure Water Monterey is intended to be a multi-benefit, integrated, regional solution that provides water recycling, improved water quality and groundwater recharge [28]. The goals are manifold. Wastewater from industrial processing in the City is diverted to a treatment facility, rather than being discharged into local waterways and eventually into the Monterey Bay. Once the water is treated, it can be used to help recharge the groundwater, preventing further saltwater intrusion and providing a more reliable and clean groundwater source for irrigation. Additionally, future phases of the project, which were recently funded by a $10 million Prop 1 grant, will divert very polluted water from the heavily polluted Reclamation Ditch and Blanco Drain, treating it and recycling back into the groundwater. While the integrated, multi-benefit project is laudable, some water managers wonder if it provides yet another free pass to let agriculture pollute into these waterbodies, now knowing they will be cleaned up further downstream.

Another more divisive event playing out in Salinas but one that could eventually lead to more coordinated efforts between water quantity and water quality and possibly between different agencies and stakeholders, is a lawsuit between an environmental organization, Monterey Coastkeeper and the Monterey County Water Resources Agency. In 2011, Monterey Coastkeeper sued the Water Resources Agency, claiming that the Agency was one of the major culprits of water pollution in Salinas but has never been regulated as such. Monterey Coastkeeper defends that the Water Resource Agency should be treated as a "waste discharger" because the Agency actively operates, withdraws and transports polluted waterways, including the Reclamation Ditch and the Blanco Drain and has inadequately managed the water quality under its jurisdiction. The Water Resources Agency, however, argues that even though it might divert or transport water for flood or groundwater recharge purposes (i.e., water quantity), it is not responsible for the pollution of that water (i.e., water quality). Instead water quality should be the responsibility of the individual polluters, most of which are exempt from permitting and under the umbrella of the Conditional Agricultural Waiver. After more than four years of litigation, in March 2015, Monterey County Superior Court Judge Thomas Wills found that the Water Resources Agency should indeed be considered a discharger. Soon after, the Agency filed a notice of intent to appeal, extending the controversy. If, in the final ruling, the court rules again in favor of regulating the County's Water Resource Agency as a discharger, while there may be unresolved questions of fairness, the decision would inevitably encourage more interagency collaboration since ultimately the County would need to work with agriculture to clean up the water they manage.

An equally important piece of the water law and policy puzzle, especially in diverse landscapes, is the set of laws and disparate agencies governing land use. Most land use activities present significant potential to pollute nearby waterways. According to at least one regional water resource regulator, land use is a significantly overlooked harm to local waterways and the agencies that control and manage land uses should be held more accountable. One of the most difficult challenges is that land use policy is fragmented among numerous, disparate entities [27]. For example, in the Salinas area, while the primary entity handling land use permits is the County of Monterey's Resource Management Agency, a variety of other entities and levels of government hold authority over planning, zoning and development patterns. Land use policies usually give little consideration to water quality and water quantity impacts.

One project on the horizon for the City of Salinas, which will test the effectiveness of transferring land uses from agriculture to less polluting activities, such as conservation and parks/recreation, is the Carr Lake project. In January 2017, The Big Sur Land Trust, a nonprofit conservation organization, bought 73 acres of seasonally dry lakebed. For decades, three Japanese farming families have owned the land that makes up Carr Lake and two continue to own and farm their parcels. The Big Sur Land Trust has goals of using the land for parks and recreation community programs, environmental initiatives and education. The City has long thought the land would be an ideal location for a "Central Park". In addition to the potential for possible water quality improvements, it also highlights the important role that nongovernmental actors and partners can and have increasingly played in helping solve a wide array of public problems, including water quality. Because public agencies are so preoccupied with budgeting, completing lengthy regulatory paperwork and trapped specializing in crucial operational functions, society becomes increasingly reliant on third party agencies to help pursue public purposes ([29]).

7. Conclusions: From Fragmented to Joint Responsibilities

"It is difficult to imagine a legal and policy regime as fractured as that used to govern water resources in the United States. Connected issues are addressed without coordination and authority is divided among federal, state and local entities that have little incentive to coordinate their interrelated actions."

—Robert W. Adler, Professor of Law

As the Salinas case study illustrates, the compartmentalized laws separating water pollution control policies in agricultural lands versus those in urban areas, as well as policies dividing water quality, water quantity and land use, do not lend themselves to coordinated and collaborative efforts—although the waterways themselves are inextricably connected. As Dr. Adler's quote eludes to, this is not an unusual case; rather it is a common phenomenon occurring throughout the United States. Managers that administer municipal separate storm sewer systems (MS4s) are using the limited resources available to them to narrowly focus on complying with their own set of permits, while other dischargers (e.g., growers) are each individually concerned with separate mandates.

While current water laws and policies tend to facilitate more divisive rather than cooperative approaches, what do seem to be driving progress are issue-based mandates and local watershed projects that occur as a result of dire circumstances. Unfortunately, these forced-collaborations are the exception to the rule and are implemented as a result of devastating events—a historic drought (SGMA), a lack of clean drinking water (Human Right to Water) and seawater intrusion (Pure Water Monterey)—rather than as a preventative measure and as standard practice. However, because of the legislative mandates' strong directives, especially the "right to consider" clause in the Human Right to Water and the SGMA landmark decision to finally regulate groundwater, each could likely set a new precedence of more interagency cooperation within California's water management network. While the SGMA and Human Right to Water are not strong enough on their own to overhaul disparate federal and state policy regimes, combined and with other local efforts, they may begin to elicit incremental and vital change.

Addressing the root cause of the pollution problem—a political system that largely exempts agriculture's pollution—would be a logic place to begin in solving this water quality conundrum.

The inadequacies of the current approach to agricultural nonpoint source pollution control in the U.S. have been widely studied and discussed (see [30] for a review of the literature). Transitioning from voluntary mechanisms to more effective and enforceable regulatory instruments based on measurable water quality performance is paramount to cleaning U.S. waters.

In California, a shift to numeric performance standards for agriculture necessitates a major overhaul of the state's Conditional Agricultural Waiver program. The most effective, yet dramatic change would be to remove the waste discharge exemption status for agriculture altogether and transition the industry from a Waiver program to a permit system. This major reform could either occur at the state or federal level. Two other nonpoint pollution sources that have successfully undergone such a transition include: (1) Confined Animal Feeding Operations (CAFOs) and (2) urban storm water.

If the California State Water Board and Regional Boards continue to work within the confines of the Conditional Agricultural Waiver program, several glaring issues need to be addressed. First, the Agricultural Waivers need to work as intended: Agricultural Waivers need to become more rigorous with every 5-year-update if water quality improvements are not being achieved and the State Board needs to hold Regional Boards accountable to making progress. While Regional Boards may have attached more provisions to their Waivers, in most regions water quality continues to decline or remains in a dismal condition with only limited success stories (e.g., water column toxicity decrease in the Central Coast region). Second, as the Superior Court Judge advised, Waivers need to have "adequate performance standards and feedback mechanisms to assess the effectiveness of implemented management practices in reducing pollution and preventing further degradation of water quality" [20]. Even if Agricultural Waivers have added modest monitoring requirements to subsequent iterations of their region's Waiver program, many still fall short of verifying the effectiveness of on-farm water quality management practices. Additionally, most programs, if not all, have insufficient monitoring data to identify individual operations that cause impairments. This issue points to a third and perhaps the most important problem within the Agricultural Waiver program: public disclosure and information. The most effective means of identifying a polluter is to conduct individual discharge monitoring at the edge of a discharger's field where pollutants enter the water. Growers and the California agricultural lobby have fought hard to keep monitoring as far away from their fields as possible and to ensure that their names are not associated with monitoring data. Growers do not want to be identified as a point source polluter and subsequently regulated under the NPDES permit system. As one Regional Board staff member put it, growers "don't want to deal with a government agency managing their land and water and they don't want to be called part of the problem". This final piece unveils the underlying impediment to agricultural water quality—as it currently functions, the Waiver program allows the most egregious agricultural polluters to hide in the shadows of collective monitoring and pollute into state waterways. Whether we transition to a permit system where more rigorous monitoring is mandated and public disclosure is law (e.g., NPDES), or whether we demand similar provisions in Agricultural Waivers, the public and policymakers must have access to transparent, sound and ongoing water quality data at the edge-of-field in order to make decisions on behalf of public health and the environment.

Another, separate agricultural-specific mandate would be to introduce a piece of legislation that addresses groundwater pollution, especially nitrate contamination. As one staff member working closely with the Salinas Valley SGMA reported, what is trying to be done with the SGMA is so colossal that the water quality piece will likely take a back seat to the focus on water quantity. A policy that would directly address nitrate contamination would be profound in many ways but perhaps one of the most important impacts it could have is in bridging water quality (especially nitrate contamination), water quantity (since water quality in groundwater is inextricably connected to groundwater levels) and land use (the biggest culprit in California being agriculture). As one groundwater expert explained, "there is no use in putting all the effort into balancing groundwater levels, if all the water is polluted and unusable" (Personal Communication, 13 November 2017). One problem cannot be solved without addressing the other, necessitating what he called a "One Water" perspective. Additionally, due to the nature of nitrogen transport pathways into groundwater, the simple process of initiating a groundwater quality regulatory program could provoke a different way of thinking about environmental regulation. This is due to the fact that regulators would need to begin monitoring and controlling the source of nitrate pollution—fertilizers—if improvements were actually to be achieved. While reducing pollutants at the point of entry might not sound revolutionary, such a policy tool has a lot of potential but is rarely been employed. Most environmental policy tools employed in the U.S. occur at the end-of-pipe, after the pollutant has already entered the environment.

The other key missing element to solving the water quality conundrum is sufficient funding. Many of the public water pollution control systems are so poorly funded that even if there were intentions to reach across jurisdictional boundaries to other dischargers, they may not have the time nor resources for undertaking such an endeavor. California's State Water Board and the Federal EPA seem to be aware of the funding problem and do offer some financial support. The State Revolving Fund, administered by the State Water Board, is one of the most common ways storm water managers secure funds for capital improvement projects. The Fund is a loan program intended to assist the following projects: nonpoint source pollution control programs, implementation of estuary conservation programs and construction of wastewater treatment facilities. Other State related funding mechanisms include Proposition 13: Safe Drinking Water Bond Act, Proposition 40: Water, Habitat, Air and Park Projects, Proposition 50: Water Security, Drinking Water, Coastal and Beach Protection Act, Federal Urban Creek Restoration Grants and State/Tribal Wetlands Program (see Salinas Stormwater Management Plan for more details). Federal grants include Nonpoint Source Implementation Grant and the Stream Restoration Mitigation Bank.

While the City of Salinas has the above grant and loan options at their disposal for special projects, their year-to-year budget for the entire storm water and urban water quality program is reliant on its portion from the General Fund. In a city plagued by high crime rates, the majority of the Salinas' General Fund budget (usually over 70%) is dedicated to funding police, fire and other public safety services, leaving the municipal storm water programs with limited staff and funding. The Public Policy Institute of California estimates an annual funding gap of $500–$800 million for storm water programs in cities throughout the state [1]. Because the State now mandates that cities comply with more rigorous NPDES permits, not only for their wastewater treatment but also for urban storm water, increased funding must follow.

Fortunately, despite the financial obstacles, Salinas has successfully complied with the most rigorous municipal discharge permit in the Central Coast region and even found creative ways to implement multi-benefit, integrated water quality and quantity projects, such as Pure Water Monterey. Nongovernmental agencies have stepped in to play a role, such as the Big Sur Land Trusts endeavors at Carr Lake, as well as third party monitoring groups aiding in assessing the regional water quality problem (e.g., Coastal Watershed Council's First Flush). Clearly, however, water quality efforts would benefit most not from disparate efforts but from larger institutional change. Currently, different levels of government are deemed as competing power structures [27]. As the urban-agricultural interface so vividly illustrates, agencies charged with regulating urban water discharges are pitted against those regulating agricultural pollution because they do not want to be responsible for each other's waste and similarly those managing water transport, flooding and recharge (e.g., Monterey County Water Resource Agency) do not want any part of water pollution management. Perhaps if the agricultural industry was held to the same high pollution standards with similar regulatory mechanisms as their urban counterparts, not only would there be a better chance of improving water quality but water resource management might have a chance of becoming integrated.

Acknowledgments: The author acknowledges financial support from the Science, Technology & Society Program of the National Science Foundation under NSF Award number 1749062.

Conflicts of Interest: The authors declare no conflict of interest.

References

1. Chappelle, C.; Hanak, E. *California's Water Quality Challenges*; Public Policy Institute of California: October, 2015; Available online: http://www.ppic.org/content/pubs/jtf/JTF_WaterQualityJTF.png (accessed on 5 November 2017).
2. National Aeronautics and Space Administration (NASA). *NASA Analysis: 11 Trillion Gallons to Replenish California Drought Losses*; NASA publication 14-333; NASA: Washington, DC, USA, 2014.
3. Goldenberg, S. *The Central Valley is Sinking: Drought Forces Farmers to Ponder the Abyss*; The Guardian: London, UK, 2015.
4. United State Environmental Protection Agency (EPA). *Assessment TMDL Tracking and ImplementatioN System (ATTAINS)*; USEPA: Washington, DC, USA, 2012.
5. California Department of Water Resources (CA DWR). Available online: www.water.ca.gov/climatechange/ (accessed on 2 November 2017).
6. Harter, T.; Lund, J.; Darber, J.; Fogg, G.; Howitt, R.; Jessoe, K.; Pettygrove, S.; Quinn, J.; Viers, J. *Addressing Nitrate in California's Drinking Water*; SBX2-1; Center for Watershed Sciences, University of California: Davis, CA, USA, 2012.
7. Knobeloch, L.; Salna, B.; Hogan, A.; Postle, J.; Anderson, H. Blue babies and nitrate contaminated well water. *Environ. Perspect.* **2000**, *108*, 675–678. [CrossRef]
8. International Human Rights Law Clinic (IHRLC). *The Human Right to Water Bill in California: An Implementation Framework for State Agencies*; University of California, Berkeley, School of Law: Berkeley, CA, USA, 2013.
9. Shilling, F.; Sommarstrom, S.; Kattelman, R.; Florsheim, J.; Henly, R.; Washburn, B. *The California Watershed Assessment Manual*; California Resource Agency: Sacramento, CA, USA, 2005.
10. United States Environmental Protection Agency (EPA). *National Water Quality Inventory*; EPA: Washington, DC, USA, 2010.
11. Andreen, W.L. Water quality today: Has the Clean Water Act Been a Success? *Ala. Law Rev.* **2004**, *55*, 537–593.
12. Anderson, B.S.; Hunt, J.W.; Phillips, B.M.; Nicely, P.A.; Vlaming, V.; Connor, V.; Richard, N.; Tjeerdema, R.S. Integrated assessment of the impacts of agricultural drainwater in the Salinas River. *Environ. Pollut.* **2003**, *124*, 523–532. [CrossRef]
13. Anderson, B.S.; Phillips, B.M.; Hunt, J.W.; Connor, W.; Richard, N.; Tjeerdema, R.S. Identifying primary stressors impacting macroinvertebrates in the Salinas River: Relative effects of pesticides and suspended particles. *Environ. Pollut.* **2006**, *141*, 402–408. [CrossRef] [PubMed]
14. Anderson, B.; Hunt, J.; Markiewics, D.; Larsen, K. *Toxicity in California Waters. Surface Water Ambient Monitoring Program*; California State Water Resources Control Board: Sacramento, CA, USA, 2011.
15. Pereira, W.; Domagalski, J.; Hostettler, F.; Brown, L.; Rapp, J. Occurrence and accumulation of pesticides and organic contaminants in river sediment, water and clam tissues from the San Joaquin River and tributaries, California. *Environ. Toxicol. Chem.* **1996**, *15*, 172–180. [CrossRef]
16. Paul, M.; Meyer, J. Streams in the urban landscape. *Ann. Rev. Ecol. Syst.* **2001**, *32*, 333–365. [CrossRef]
17. Roy, A.H.; Wenger, S.H.; Fletcher, T.D.; Walsh, C.J.; Ladson, A.R.; Shuster, W.D.; Thurston, H.W.; Brown, R.R. Impediments and solutions to sustainable, watershed-scale urban stormwater management: Lessons from Australia and the United States. *Environ. Manag.* **2008**, *42*, 344–359. [CrossRef] [PubMed]
18. United States Environmental Protection Agency (EPA). *National Pollutant Discharge Elimination System (NPDES): Sanitary Sewer Overflows (SSOs)*; EPA: Washington, DC, USA, 2016.
19. Otter Project; Monterey Coastkeeper; PCFFA; Environmental Justice Coalition for Water; Santa Barbara Channelkeeper; Sportfishing Protection Alliance. "Judge Rules State Must Do Much More to Curb Agricultural Pollution." (Press Release). Available online: http://mavensnotebook.com/wp-content/uploads/2015/08/Press-Release-Ag-Order-Win.png (accessed on 1 December 2017).
20. Monterey Coastkeeper; The Otter Project; PCFFA; Environmental Justice Coalition for Water; Santa Barbara Channelkeeper; Sportfishing Protection Alliance v. *California State Water Resources Control Board (SWRCB)*; Case Number: 34-2012-80001324; Superior Court of California, County of Sacramento: Sacramento, CA, USA, 2015.
21. Meurer, F. *Letter to the Regional Board Posted in: Transcripts of Proceedings*; Meeting minutes; Central Coast Regional Water Quality Control Board: San Luis Obispo, CA, USA, 2011.

22. Poteete, A.R. Levels, scales, linkages, and other 'multiples' affecting natural resources. *Int. J. Commons.* **2012**, *6*, 134–150. [CrossRef]

23. City of Salinas. Stormwater Management Plan Update. 2013. Available online: https://www.cityofsalinas.org/our-city-services/public-works/water-waste-energy/stormwaterwater-recycling/stormwater-documents-3 (accessed on 1 November 2017).

24. City of Salinas Municipal Storm Water Discharges Technical Report. 2012. Available online: https://www.waterboards.ca.gov/rwqcb3/water_issues/programs/stormwater/docs/salinas/2012_0005_salinas_fact_sheet.png (accessed on 14 November 2017).

25. Central Coast Ambient Monitoring Program (CCAMP). Available online: http://www.ccamp.org/ (accessed on 20 November 2017).

26. Surface Ambient Monitoring Program (SWAMP). Available online: https://www.waterboards.ca.gov/water_issues/programs/swamp/ (accessed on 10 November 2017).

27. Adler, R.W.; Straube, M. *Watersheds and the Integration of U.S. Water Law and Policy: Bridging the Great Divides*; William & Mary Environmental Law Policy Review: Williamsburg, VA, USA, 2010; Volume 25.

28. Pure Water Monterey. Available online: http://purewatermonterey.org/ (accessed on 13 November 2017).

29. Salamon, L.M. The new governance and the tools of public action: An introduction. In *Tools of Government: A Guide to the New Governance*; Oxford University Press: New York, NY, USA, 2002; pp. 1–47.

30. Drevno, A. Policy Tools for Agricultural Nonpoint Source Water Pollution Control in the U.S. and E.U. *J. Manag. Environ. Qual.* **2016**, *37*, 106–123. [CrossRef]

resources

MDPI

Communication

California's Groundwater Regime: The Cadiz Case

Julia Sizek

Anthropology Department, University of California-Berkeley, Berkeley, CA, 94720, USA; jsizek@berkeley.edu

Received: 18 November 2017; Accepted: 18 January 2018; Published: 21 January 2018

Abstract: Recent California legislation has promised solutions to longstanding problems in groundwater management through an emphasis on management of groundwater itself, rather than on the rights of overlying property owners. In this short communication, I argue that the promises of scientific management relies on property law and jurisdiction and therefore that scientific claims about the water itself are less important than private property claims in the case of a Cadiz Inc.'s proposed groundwater extraction project in Southeastern California. While private property in land insulates Cadiz Inc. (Los Angeles, CA, USA) from political contestation, opposition to the project has increasingly focused on the right to transport and transfer water through lands not held by Cadiz Inc. This legal strategy points to how California groundwater law is still fundamentally ruled by private property in land, which shifts the grounds of environmental politics from extraction itself to the transport of extracted materials. This case serves as a good example of the intersection of political ecology and legal geography.

Keywords: legal geography; groundwater; private property; political ecology; water rights; right-of-way; scale

1. Introduction

California depends on groundwater: 50% of water used in the state comes from underground stores. Despite its importance to water provision throughout the state, groundwater has remained largely invisible to the general public. As a result, groundwater-related problems such as overdraft and land subsidence plague the state [1].

The 2014 passage of the Sustainable Groundwater Management Act (SGMA) promises to solve California's recalcitrant groundwater problems by creating agencies to oversee groundwater that correspond to groundwater basins. Prior to SGMA, California's groundwater was determined on a case-by-case basis as landowners overlying the same aquifer sued each other to adjudicate their rights to extract water from the basin below [2]. The case-by-case adjudication resulted in a patchwork of legal regulation around the state, which has been hailed by some as successful instances of polycentric governance structures and disparaged by others as piecemeal policy [2,3]. The passage of SGMA ensured groundwater plans would be made for medium- and high-priority basins where adjudication had never occurred between overlying property owners.

Basin-scale management has long been the dream of California's water managers, and the idea of watershed-scale management dates back to John Wesley Powell's explorations of the American West in the mid-1800s [4,5]. Matching management scale to hydrologic scale promises to reshape groundwater management from property boundaries to scientific ones rather than the politics of overlying owners [5].

This short communication examines this central promise of SGMA—that groundwater itself can be centered in debates about proposed extraction, rather than the property rights of overlying owners. Through examining a legal contestation and public discourse around a proposed groundwater project in Southeastern California, this short communication argues that California groundwater policy is ruled by private property rights, not by scientific management practices. This fact shifts possibilities for legal contestation away from the project itself and toward the material infrastructure needed to complete groundwater transfer. In the first section, I describe how the Cadiz Inc. project fits into the history of California water and private property law pre-SGMA, introducing the key concepts of "reasonable use" and "surplus water" as key ways of understanding the relationship between water and property. In the second, I describe challenges to the Cadiz Inc. project in the form of scientific questions about groundwater pumping rates and their lack of success in creating meaningful debate around the project because jurisdictional issues raise the question not of whether the project should happen but rather of how much they are allowed to pump. Finally, I discuss how groundwater transport has become the primary site of contestation over the project, and the ways that this reinforces private property regimes.

2. Background: The Cadiz Project under California Groundwater Law

Although Cadiz Inc. has often appeared to be an agricultural company, their 1987 SEC 10-K filing indicates that the original incorporated company, Aridtech, was a "water exploration and development company" meant to acquire underground water resources. In 1985, Aridtech incorporated the Cadiz Valley Development Incorporation, and both of these organizations became the Cadiz Inc. Company known today. Over time, Cadiz Inc. has added more land, and now they have approximately 40,000 acres in the Eastern Mojave around their project area, though the land is not contiguous [6].

In 1983, the company now known as Cadiz Inc., a water and natural resources company, acquired about 14,000 acres of creosote scrub in the Eastern Mojave Desert on the assumption that the relatively closed aquifer underneath it would yield profits once the water was moved and sold [6]. Like other agricultural speculators after the 1977–1978 drought, they sought to take advantage of the property tax benefits of Proposition 13 and low land prices to invest in the land market. This investment was based on the idea that acquiring overlying land would grant rights to water that could appreciate in value [7]. Later, the groundwater could be transferred to a new owner because of 1982 laws that eased groundwater permit transfers to private and public water companies [8]. In the 1990s, the company unsuccessfully proposed a groundwater storage project (similar to the Kern Valley Water Bank), and today's iteration of the project is to extract and transport it to coastal water districts. This project is controversial, and almost universally derided by those living in the Mojave Desert, where the project is located. In more than 20 interviews I conducted in the summer and fall of 2017 with residents of the Morongo Basin, East Mojave Desert, and Barstow, most locals adhered to a similar narrative: they oppose the project as a profitable "water grab" for Cadiz Inc. that would subsidize irresponsible water use by coastal Californians to the detriment of a fragile desert ecology.

Though local opposition to the project is relatively high, the project has legal precedent in California. Since no statewide oversight for groundwater existed until 2014 [9], groundwater before that time was governed by a legal precedent known as "California correlative rights," or "reasonable use," which is distinct from the laws governing surface water in the state. California's correlative rights doctrine came from the 1903 court case *Katz v. Walkinshaw*, which established that overlying property owners had access to the water below and that such limits to the use of water could only be proven if they extended beyond "reasonable use" [10,11]. Though these water rights are still use rights rather than property rights, California's system is only one step removed from treating water as absolute private property, as in Texas [12].

"Reasonable use" creates an interesting problem in the relationship between water and land in California groundwater law: though it is primarily used as a way to limit overlying users from using too much water or wasting water, it makes possible the legal phenomenon of "surplus" water [8]. "Surplus" water is created when the overlying owners' reasonable use is determined, but that the combined amount of water needed by each user is less than the total water that can be safely extracted each year [10]. This so-called surplus water can then be appropriated and conveyed to non-overlying users through groundwater permitting schemes, largely because the California Constitution mandates beneficial use of water by humans over in-stream uses for wildlife [13].

For the Cadiz project, the creation of "surplus" water in the aquifer is central to their profit-making strategy. Unlike many overdrafted aquifers throughout the state shared between private landowners, the aquifer underlying the Cadiz, Fenner, and Bristol Valleys are shared largely between Cadiz Inc. and the federal government. In this area, the federal government has not sued to adjudicate the basin and seems to favor human use over in-stream uses for local flora and fauna [6,14]. As a result, all appropriable groundwater can go to Cadiz Inc. even if their use extends beyond reasonable use needed for their lands.

The California legal system has created a unique legal ecology in which projects like Cadiz can thrive by basing access to groundwater on the private rights of overlying landowners. Like mineral rights, these water rights are both connected to and alienable from the land under property law [15]. (On the limits of water-as-mineral, see *Andrus v. Charlestone Stone Prods. Co.*, (436 U.S. 604, 1978), a Supreme Court case heard in which a man tried to file a patent under the 1872 mining act for the water underneath the land. In the decision, the Supreme Court determined that groundwater could not be patented under the act because it was neither named in the act nor under the jurisdiction of the federal government). The dual legal status of water rights also demonstrate how critical legal geography—understanding the legal underpinnings of resource extraction—intersects with the political ecology of extraction and profit [16–18]. In the following two sections, I focus on the ways that environmental organizations and others have contested the Cadiz project through tracing legal action surrounding the Cadiz project. Through following these legal questions, I argue that scientific basin-wide claims about water have been legally unsuccessful, as claims about private property in conveyance areas have had greater traction, revealing how property law rules water rights.

3. The Contested Science of the Groundwater Basin

Although groundwater is an extremely important component of water provision, it is known to many managers as the "unseen" or "invisible" resource because it is hidden from sight and its importance remains unrecognized by the general public [19]. While groundwater is increasingly important in providing water to consumers (especially in times of drought), much remains uncertain about the groundwater hydrology. In the Cadiz case, both Cadiz Inc. and environmentalist opposition have used this uncertainty to make claims about the effects of the project during the California Environmental Quality Act (CEQA) process. Opposition to the project has proven unsuccessful at challenging the hydrological basis and jurisdiction of the project, demonstrating how only the extent of the project can be questioned, not its validity.

Two particularly contested issues are basin connectivity and recharge rates, both of which are central to understanding the long-term effects of groundwater extraction, which include effects to local water sources, saltwater intrusion in coastal areas, and land subsidence. Aquifers are most often imagined as closed bowls containing water, an image that serves a utilitarian and political purpose in delineating communities of interest [5,20]. Few aquifers fulfill this ideal. Subsurface geology determines where the aquifer connects to other underground water sources (in California, this has also appeared famously in debates about connectivity between surface water and groundwater, in *City of Los Angeles v. Pomeroy* (1899)), or in what ways overdraft from that aquifer could affect surrounding water sources. In the case of the Cadiz project, the limits of the aquifer are hotly contested: environmental groups have claimed hydraulic connection between the Bonanza Spring, a year-round water source for desert flora and fauna, and the Fenner and Orange Blossom aquifers, though the company denies this connection. Even if such beneficial use of the water were proven, it would be unlikely to hinder the project: indirect benefits of water to wildlife are typically understood to be less important than direct beneficial use to people [14].

Calculating groundwater recharge rates is also notoriously complex, since it depends on the collection of rainwater that percolates to the aquifer (and possibly other stream or river sources). Recharge rates are also affected by discharge rates, including springs and dry lakebeds, both of which are at issue for this aquifer [20]. Though these processes are simple in concept, they are difficult to measure practically, and many proposed recharge rates vary widely. In the case of the Cadiz project, groundwater recharge rates range from a 2000 Maxey/Eakin study by USGS predicting 2550 to 11,200 acre-feet per year to a Cadiz-funded study in 2010 that gave a proposed recharge rate of 32,000 acre-feet per year [6,21]. (Of note: The scientific analyses for the project propose that the water transferred to SMWD would be otherwise lost to evaporation through dry lake beds, so the project "saves" water from natural inefficiencies, using the same logic of the "wastefulness" of nature outlined in Gidwani and Reddy's (2011) analysis of the simultaneously moral and economic character of languages of water use [22]). While these differences in estimates could be quite consequential, they are hard to assess when the project is enacted, since many detrimental impacts of pumping are only seen long after the pumping is completed [20].

Much of the science surrounding the Cadiz aquifer has been contested by environmental agencies and by the federal government during the CEQA process, as attested above. Because so much of CEQA is based on mitigating harms rather than eliminating them [22], these challenges have not fundamentally questioned Cadiz Inc.'s right to draw so-called "surplus" water from the basin beneath them and instead have only questioned the quantity of water available to Cadiz. The reason why these legal challenges by environmentalists have failed is not because of the invalidity of scientific questions, but because of jurisdiction and private property: California's correlative rights doctrine has insulated Cadiz Inc. from legal challenges about their right to move water beneath their land. The question as framed under CEQA, like the technical solutions under SGMA, is one of *how much* extraction is permissible, not *whether* such extraction should be allowed. As a result, the legal arena for contestation about the Cadiz project has shifted to the actual conveyance of water to SMWD through an old railroad right-of-way. This right-of-way, not the aquifer itself, has been the most fruitful site of legal contestation over the project.

4. Pipeline Politics

Cadiz Inc. has long planned to transfer water away from their pumping stations through a pipeline they would construct in the right-of-way on the Arizona and California railroad line that travels from Cadiz to the Colorado River aqueduct. As much as the public discourse surrounding the water transfer has generally played out along the ethics of selling and transporting water from the desert to coastal Orange County (as seen in my interviews with members of the public, newspaper articles [23], and op-eds [24,25]), the most concrete legal action has been about the property regimes of the transport process, and whether Cadiz Inc. should be allowed to use old railroad right-of-ways without federal review. The success of this legal strategy in opposing the project points to the underlying importance of property in land in determining the fate of the project.

Built from 1905 to 1907, the Arizona–California line of the Santa Fe railroad traveled from Cadiz (CA) to Parker (AZ), receiving a right-of-way from the federal government, as did many transcontinental railroads of the time [26]. This grant, under the 1875 railroad right-of-way act, is between an easement and an outright fee grant of the land, as had been determined by *Great Northern Railway Company v. United States* (1942). The status of railroad right-of-ways had always been a moot point, and was debated in myriad forms in *Leo Sheep Co. v. United States* (1979) and *Home on the Range v. AT&T Corporation* (2005), which determined that (1) grants from the federal government are "construed strictly against the grantees" and (2) that railway right-of-way easements were limited to that of a railroad purpose following the narrow interpretation of the easement.

Under an Obama-era Department of the Interior solicitor's decision (M-37025), Cadiz Inc. would have to undergo federal review to build their pipeline on federal lands because the proposed pipe did not further a railway purpose under *Home on the Range* and therefore did not fall within the grant provided to the railway company, narrowly construed. This 2011 decision decided on the basis of property law that Cadiz Inc.'s conveyance pipeline would have to undergo federal review (rather than only CEQA, which had dictated the terms of their previous review). However, this decision was rescinded and replaced with M-37048 by the Trump Administration's Acting Solicitor Daniel Jorjani in

September 2017. In this new decision, Jorjani argues against the Supreme Court's decision in *Home on the Range*, instead stating that railroad right-of-ways should be understood as a gross easement with divisible bundles of rights that can be sold at a profit (as in property, cf. Kay [27]). Under his guidance, the railroad maintains the right to divide up its right-of-way easement as long as the new right-of-way use does interfere with railroad activities.

As a rejoinder to the Trump Administration's efforts to have the project not receive further review, the state of California has found that it owns lands underlying the railroad right-of-way and therefore that the California State Lands Commission will have to review the proposed project conveyance for a lease. At the time of writing, decisions are not final as to what the California State Lands Commission will decide in their review process. However, what both of these cases show is how the politics of groundwater are fought not through the water itself or the right to extract, but through the land above it—in the latter case, the land through which the water must be conveyed. This points toward the ways that questions of private property in land—even outside of the area properly understood as part of the basin—come to determine groundwater politics.

What is so legally contested in the case of right-of-ways is exactly the discussion *not* being asked in the case of overlying ownership of groundwater: what is the degree to which a right-of-way (a form of use-right) conveys a right in property? This litigation reveals a re-working of legal scale to question the boundaries of property and rights ownership [28] and to reconfigure the ways in which property rights are important for understanding contemporary extractive projects.

5. Conclusion: Property in Water

Property law is both the central tool Cadiz Inc. has in protecting their groundwater extraction plans and their chief enemy as they attempt to convey the water through the desert to the Colorado River Aqueduct. The legal debates that have emerged from Cadiz Inc.'s proposed conveyance are in exact opposition to those that SGMA conceptually relies on: rather than bringing parties to the table to discuss aquifer recharge rates and the science of groundwater, the debate of the project is about a distinction between property and right-of-way. The pipeline infrastructure becomes the means of debating the entirety of the project, as in the case of debates over Keystone XL and Dakota Access pipelines [29,30]. Unlike these fossil fuel pipelines, Cadiz's proposed pipeline does not pose a danger to people—in fact, Cadiz Inc. has attempted to permit the pipeline as necessary for railroad fire suppression—but the underlying debate is about whether the project will harm wildlife by extracting groundwater and whether it is appropriate to remove water from the desert for profit. The debates over the right-of-way are a proxy for debates made impossible under the legal system: should water be sold for profit? Do affected wildlife have rights?

The railroad debate combines aspects of the political economy of water extraction and legal scale through revealing the private property rights that both promote water extraction and make water conveyance difficult. In this intersection, the controversial groundwater project faces its most potent opposition through the pipeline rather than through claims on the water rights of the overlying property owners themselves, much like in cases of fossil fuel industries. The politics of the groundwater project are not seen through the still-invisible aquifer, but instead through the visible pipeline that will travel through the California desert.

Acknowledgments: This research has been supported by a fellowship from the National Science Foundation's Graduate Research Fellowship Program.

Conflicts of Interest: The authors declare no conflict of interest. The funding sponsors had no role in the design of the study; in the collection, analyses, or interpretation of data; in the writing of the manuscript; or in the decision to publish the results.

References

1. Faunt, C.; Sneed, M.; Traum, J.A.; Brandt, J. Water Availability and Land Subsidence in the Central Valley, California, USA. *Hydrogeol. J.* **2016**, *24*, 275–284. [CrossRef]

2.   Blomquist, W.A. *Dividing the Waters: Governing Groundwater in Southern California*; ICS Press: San Francisco, CA, USA, 1992.

3.   Tarlock, A.D. Putting rivers back in the landscape: The revival of watershed management in the United States. *Hastings West Northwest J. Environ. Law Policy* **2000**, *6*, 167–195.

4.   Powell, J.W. *Report of the Lands of the Arid Region of the United States, With a More Detailed Account of the Lands of Utah*, 2nd ed.; Government Printing Office: Washington, DC, USA, 1879.

5.   Blomquist, W.; Schlager, E. Political Pitfalls of Integrated Watershed Management. *Soc. Nat. Resour.* **2005**, *18*, 101–117. [CrossRef]

6.   Santa Margarita Water District. *Cadiz Valley Water Conservation, Recovery, and Storage Project*; Final Environmental Impact Report; SCH# 2011031002; Santa Margarita Water District: Rancho Santa Margarita, CA, USA, 2012.

7.   Gaffney, M. What Price Water Marketing?: California's New Frontier. *Am. J. Econ. Sociol.* **1997**, *56*, 475–520. [CrossRef]

8.   Gray, B.E. A Primer in California Water Transfer Law. *Ariz. Law Rev.* **1989**, *31*, 745–781.

9.   Escriva-Bou, A.; McCann, H.; Hanak, E.; Lund, J.; Gray, B. *Accounting for California's Water*; Public Policy Institute of California: San Francisco, CA, USA, 2016.

10.   Brown, J.A. Uncertainty Below: A Deeper Look into California Groundwater Law. *Environ. Law Policy J.* **2015**, *39*, 45–95.

11.   Sax, J.L. We Don't Do Groundwater: A Morsel of California Legal History. *Univ. Denver Water Law Rev.* **2003**, *6*, 270–317.

12.   Gardner, R.; Moore, M.R.; Walker, J.M. Governing a Groundwater Commons: A Strategic and Laboratory Analysis of Western Water Law. *Econ. Inq.* **1997**, *35*, 218–234. [CrossRef]

13.   Cantor, A. Material, Political, and Biopolitical Dimensions of 'Waste' in California Water Law. *Antipode* **2017**, *49*, 1204–1222. [CrossRef]

14.   Losi, C.J. Keeping Dry Streams Green: Can Landowners in Arizona and California Use Property Rights to Maintain Groundwater-Dependent Riparian Habitat Along Non-Perennial Watercourses. *Hastings West Northwest J. Environ. Law Policy* **2012**, *18*, 121–156.

15.   Blomley, N. Law, Property, and the Geography of Violence: The Frontier, the Survey, and the Grid. *Ann. Assoc. Am. Geogr.* **2003**, *93*, 121–141. [CrossRef]

16.   Andrews, E.; McCarthy, J. Scale, Shale, and the State: Political Ecologies and Legal Geographies of Shale Gas Development in Pennsylvania. *J. Environ. Stud. Sci.* **2014**, *4*, 7–16. [CrossRef]

17.   Bebbington, A. Underground Political Ecologies: The Second Annual Lecture of the Cultural and Political Ecology Specialty Group of the Association of American Geographers. *Geoforum* **2012**, *43*, 1152–1162. [CrossRef]

18.   Huber, M.T.; Emel, J. Fixed Minerals, Scalar Politics: The Weight of Scale in Conflicts over the '1872 Mining Law' in the United States. *Environ. Plan. A* **2009**, *41*, 371–388. [CrossRef]

19.   Blomquist, W.; Ingram, H.M. Boundaries Seen and Unseen. *Water Int.* **2003**, *28*, 162–169. [CrossRef]

20.   Heath, R.C. *Basic Ground-Water Hydrology*; U.S. Geological Survey: Reston, VA, USA, 1983.

21.   Dubois, S.R. *Re: National Park Service Comments to Draft Environmental Impact Report for the Cadiz Valley Water Conservation, Recovery and Storage Project*; United States Department of the Interior: Washington, DC, USA, 2012.

22.   Gidwani, V.; Reddy, R.N. The Afterlives of 'Waste': Notes from India for a Minor History of Capitalist Surplus. *Antipode* **2011**, *43*, 1625–1658. [CrossRef]

23.   Aron, H. *A Controversial Plan to Drain Water from the Desert? Go for It, Trump Administration Says*; L.A. Weekly: Culver City, CA, USA, 2017.

24.   Feinstein, D.; Friedman, L. *The Scheme to Pump Desert Water to L.A. Could Destroy the Mojave. California's Legislature Needs to Block It*; Los Angeles Times: Los Angeles, CA, USA, 2017.

25.   Castro, L. *Protect the Groundwater beneath Our National Treasures*; SCV: Santa Clarita, CA, USA, 2017.

26.   Myrick, D.F. *Railroads of Nevada and Eastern California*; Howell-North Books: Berkeley, CA, USA, 1963.

27.   Kay, K. Breaking the Bundle of Rights: Conservation Easements and the Legal Geographies of Individuating Nature. *Environ. Plan. A* **2016**, *48*, 504–522. [CrossRef]

28.   Jepson, W. Claiming Space, Claiming Water: Contested Legal Geographies of Water in South Texas. *Ann. Assoc. Am. Geogr.* **2012**, *102*, 614–631. [CrossRef]

29. Barry, A. *Material Politics: Disputes along the Pipeline*; Wiley: Oxford, UK, 2013.
30. Bond, D. The Promising Predicament of the Keystone XL Pipeline. *Anthropol. Now* **2015**, *7*, 20–28. [CrossRef]

MDPI

St. Alban-Anlage 66

4052 Basel, Switzerland

Tel. +41 61 683 77 34

Fax +41 61 302 89 18

http://www.mdpi.com

Resources Editorial Office

E-mail: resources@mdpi.com

http://www.mdpi.com/journal/resources

www.ingramcontent.com/pod-product-compliance
Lightning Source LLC
Chambersburg PA
CBHW051911210326
41597CB00033B/6106